WORKBOOK FOR
Modern
Residential Wiring

by

N. Henke-Konopasek
Senior Editor
Griffith, Indiana

and

H. N. Holzman
Master Electrician
Member, International Association of
Electrical Inspectors

Publisher
THE GOODHEART-WILLCOX COMPANY, INC.
Tinley Park, Illinois

*Tools and materials for cover photography,
courtesy of Johnson's Metro Hardware Center,
Kankakee, Illinois.*

Introduction

This Workbook is designed for use with the text, **Modern Residential Wiring.** As you complete the questions and problems in this Workbook, you can refer to the facts and concepts presented in the text.

The chapters in this Workbook correspond to the chapters in the text. After reading a chapter in the text, do your best to complete the Workbook questions and problems carefully and accurately.

Each chapter of the Workbook includes objectives and instructions. Several types of questions and problems are given in each chapter. The various types of questions include identification, true or false, multiple choice, fill-in-the-blank, and written answer.

Reading **Modern Residential Wiring** and using this Workbook will provide you with a solid background of electrical wiring principles and common practices as well as a basic understanding of *National Electrical Code*® requirements. Once having mastered this information, you will be prepared to further increase your knowledge of wiring methods through additional readings and practical experiences.

N. Henke-Konopasek

H. N. Holzman

Contents

1
Electrical Energy Fundamentals

CTIVE: You will be able to explain what electricity is, how it is produced, transmitted, and measured.

CTIONS: Carefully read Chapter 1 of the text. Then complete the following questions and problems.

_____ is a form of energy that will operate appliances or produce light and heat.

According to the _____ theory, all matter is made up of atoms.

The figure below shows a typical _____ which is made up of a nucleus of protons and neutrons and orbiting electrons.

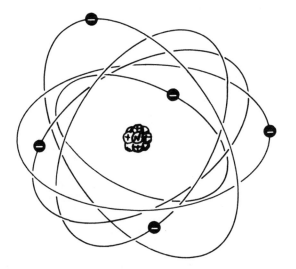

Indicate the charge of each of the following particles by matching them correctly.

_____ Protons   a. Non-charged

_____ Neutrons   b. Positively charged

_____ Electrons   c. Negatively charged

Which of the following is an example of a good conductor of electrons?

a. Glass.

b. Pure metal.

c. Dry gasses.

d. All of the above.

6. Which of the following is an example of a good insulator?

   a. Water.

   b. Carbon.

   c. Rubber.

   d. All of the above.

7. Electrons will flow from a negatively charged body to a positively charged one. This flow is called _____.

8. In electricity, differences in potential are measured in _____.

9. List the two types of electric current.

   _____

   _____

10. In _____ current, the electricity flows in one direction only.

11. Because _____ current reverses its direction, usually 60 times each second, it is called 60-cycle electricity.

12. A _____ is one complete electrical wave or vibration.

13. Energy may be produced chemically, as with a _____, or produced mechanically, as with a _____.

14. A basic chemical device for providing electrical power is the:

   a. Cathode.

   b. Anode.

   c. Cell.

   d. None of the above.

15. In a battery, the pole that is positively charged is the _____.

16. In a battery, the pole that is negatively charged is the _____.

17. Explain how a generator operates.

   _____

   _____

   _____

   _____

   _____

18. In electrical wiring, the pathway is called a _____.

19. Which of the following is necessary for a simple electrical circuit?

    a. A power source.

    b. Conductors.

    c. A load or loads.

    d. A device or devices for controlling current.

    e. All of the above.

20. List the three types of electrical circuits.

    _____

    _____

    _____

21. A _____ circuit is one in which only one path is provided for the current.

22. A _____ circuit has more than one path for the current.

23. Label the types of circuits represented by the diagrams below.

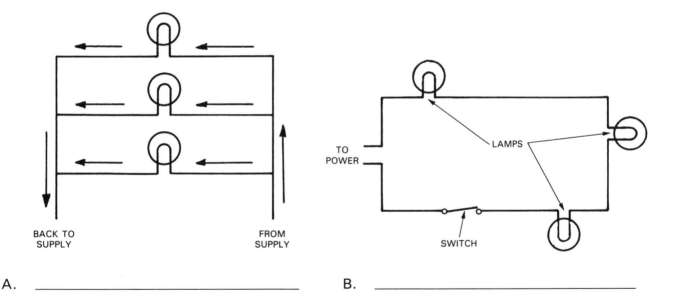

A. _____   B. _____

24. Series circuitry is practical for residential wiring. (true or false)

25. With a parallel circuit, if one load, such as a lamp, burns out, the other branches would continue to operate since a path still exists from one supply terminal through the circuit to the other supply terminal. (true or false)

26. Switches and fuses or circuit breakers are examples of residential wiring done in parallel. (true or false)

27. In electrical wiring, a _____ is any device that uses an electric current and converts the energy to another form.

28. List four examples of types of devices designed to operate at definite voltages.

_____

_____

_____

_____

29. Explain how electromagnetic induction is accomplished.

_____

_____

_____

_____

30. List the four factors that determine the voltage obtained from a generator.

_____

_____

_____

_____

31. _____ is at its greatest when the coil is moving at right angles to the magnetic field lines of force.

32. The ac generator is usually called an alternator. (true or false)

33. Explain the left hand rule for generators, also known as Fleming's Rule.

_____

_____

_____

_____

34. The parts of the alternating current generator are listed below. Match them with their descriptions on the right.

_____ Coil or armature

_____ Nonmoving poles or opposite ends of field magnets

_____ Slip rings

_____ Brushes

a. These metal parts are always in contact with one of the terminals of the coil or armature. They transfer the current to the brushes.

b. This part rotates. It has the conducting wire which cuts across a magnetic field. The electron flow begins in this part.

c. Two of these transfer current from the slip rings to the external circuit. One is in contact with each slip ring.

d. These create the magnetic field.

Label the basic parts of the alternator shown below.

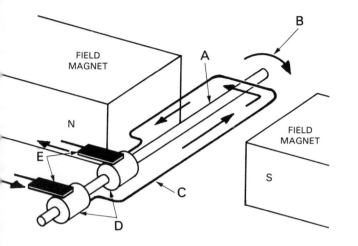

A. _____

B. _____

C. _____

D. _____

E. _____

In the United States, alternating current is _____ hertz (Hz).

a. 2.

b. 30.

c. 60.

d. 120.

The dc generator is similar in construction to the alternator except that the two slip rings are replaced by one slip ring or _____.

Label the basic parts of the dc generator shown below.

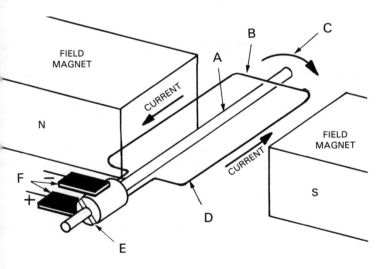

A. _____

B. _____

C. _____

D. _____

E. _____

F. _____

39. An ac generator may produce either single phase or three phase electric power. (true or false)

40. Household power is produced as _____ (single or three) phase current.

41. In industry, where heavy use is made of electricity, _____ (single or three) phase current is used.

42. Three phase motors are simpler, less expensive, and more powerful than single phase motors. (true or false)

43. Before electric power is brought into households, it is reduced to:
    a. 120 and 240 volts.
    b. 132,000 volts.
    c. 300 and 600 volts.
    d. 750,000 volts.

44. A _____ uses electric induction principles to increase or decrease voltage between an electrical power source and its load.

45. Voltage into and out of a transformer is related to the number of _____ of wire in the coils.

46. In a step-up transformer, the number of turns is greater in the secondary coil than in the primary coil, therefore output voltage will be _____ the input voltage.
    a. Less than.
    b. Greater than.
    c. The same as.
    d. None of the above.

47. In a step-down transformer, the secondary coil has fewer turns than the primary coil, therefore input voltage will be _____ the output voltage.
    a. Less than.
    b. Greater than.
    c. The same as.
    d. None of the above.

48. Explain the difference between a generator and an electric motor.

    _____

    _____

    _____

49. Label the parts of the dc motor shown below.

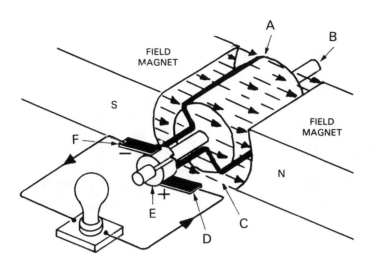

A. _____

B. _____

C. _____

D. _____

E. _____

F. _____

50. Which of the following terms or units are used to measure electricity?

    a. Amperage.

    b. Voltage.

    c. Resistance.

    d. Wattage.

    e. All of the above.

51. The rate at which electricity flows is called the _____. These units are called _____ or _____.

52. Electrical pressure or force by which electrons are moved through a conductor is termed _____. These units of measuring electrical pressure are called _____.

53. The opposition to the flow of electrons through a conductor is called _____ and is similar to friction. This is measured in units of _____.

54. The amount of power derived from an electrical device or system is called _____. This is measured in units called _____, _____, or _____.

55. Which of the following is correct?

    a. VOLTS = WATTS × AMPS

    b. AMPS = VOLTS × WATTS

    c. WATTS = VOLTS × AMPS

    d. None of the above.

56. One horsepower or 550 foot-pounds/sec. is equivalent to _____ watts.

57. The _____ _____ _____ establishes a set of rules, regulations, and criteria for the installation of electrical equipment.

58. Although the NEC, itself, has no legal basis, it is often made mandatory under local or state rulings. In such cases, if becomes a legal document. (true or false)

59. Identify and briefly describe the following.

A. _____ B. _____
   _____   _____
   _____   _____
   _____   _____
   _____   _____

      C. _____
         _____
         _____
         _____
         _____

60. The abbreviation OSHA stands for _____ _____ and _____
    _____.

61. OSHA safety regulations are often more detailed than those of the NEC and, in some instances, supercede the *Code* rulings. (true or false)

62. All persons involved in the electrical trade should become familiar with OSHA electrical standards. List the eight areas to which these standards apply.

    _____
    _____
    _____
    _____
    _____
    _____
    _____
    _____

2
Electrical Circuit Theory

e _____   Score _____

_____ Period _____   Instructor _____

:CTIVE: You will be able to apply and use electrical circuit theory.

CTIONS: Carefully read Chapter 2 of the text. Then complete the following questions and problems.

Explain what is meant by Ohm's law.

_____

_____

_____

_____

_____

_____

Mathematically, Ohm's law takes on which form?

a. $E = \dfrac{I}{R}$ c. $I = \dfrac{E}{R}$

b. $I = \dfrac{R}{E}$ d. $R = \dfrac{E}{I}$

Name two metals widely used in electrical wiring.

_____

_____

Explain why substances such as cotton, rubber, and plastic are extremely poor conductors of electricity.

_____

_____

Explain why rubber and plastic are often used as coverings for electrical wiring.

_____

_____

The longer the electrical conductor, the _____ (greater, less) its resistance to current will be.

7. Generally, the thicker the wire, the _____ (less, more) the resistance will be.

8. A _____ _____ is the cross-sectional area of a wire 1 mil or .001 in. in diameter.

9. Compute the area of a .400 in. wire in circular mils. (Show your work in the space provided.)
_____

10. As the temperature of metals rises, the resistance to current:
    a. Increases.
    b. Decreases.
    c. Stays the same.

11. List five rules that apply to series circuits.

    _____

    _____

    _____

    _____

    _____

12. Use this series circuit to answer the questions below. (Show your work in the spaces provided.)

    a. What is the total resistance in this series circuit?
       _____ ohms

    b. What is the amperage at the source, I_S, in this series circuit?
       _____ amps

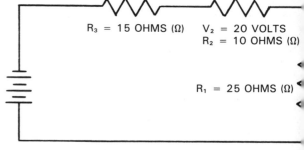

R_3 = 15 OHMS (Ω) V_2 = 20 VOLTS
                              R_2 = 10 OHMS (Ω)

R_1 = 25 OHMS (Ω)

    c. What are the voltages of resistors V_1 and V_3 in this series circuit?
       V_1 = _____ V
       V_3 = _____ V

14

3. In electricity, the term ''parallel'' means:

   a. Physically parallel.

   b. Geometrically parallel.

   c. Alternate routes.

   d. None of the above.

4. In the _____ circuit below, the arrows indicate the _____ flow. In residential wiring, these routes are called _____.

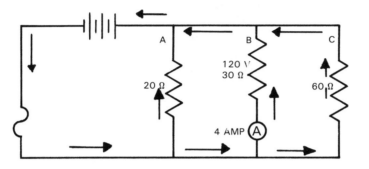

5. List five rules that apply to parallel circuits.

   _____
   _____
   _____
   _____
   _____
   _____
   _____
   _____
   _____
   _____
   _____
   _____

6. What is the total amperage of a parallel circuit having a total resistance of 15 ohms and a source voltage of 30 volts? (Show your work below.)

   _____ amps

7. When resistances are arranged both in series and in parallel, the circuit is called a _____.

8. To find the total resistance in a network circuit, reduce the parallel resistance into an _____ _____ _____ and add this resistance to the resistances arranged in series.

19. Considered as a whole, electrical circuits are mostly of the _____ type.
    a. Network.
    b. Series.
    c. Parallel.
    d. None of the above.

20. The drawing below is a way to remember the mathematical relationships of _____ _____.

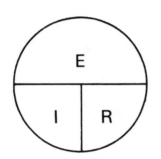

21. Use the drawing in #20 to choose the correct statement(s) below.
    a. To find E, multiply I and R.
    b. To find E, divide I by R.
    c. To find R, multiply E and I.
    d. To find R, divide E by I.
    e. To find I, multiply E and R.
    f. To find I, divide E by R.

22. Electrical energy can be measured in units called _____.

23. In electricity, power is measured in _____ or _____-_____.

24. A kilowatt-hour is equivalent to _____ being used for a 1 hour period.
    a. 100 W.
    b. 1000 W.
    c. 1,000,000 W.
    d. None of the above.

25. Which of the following would consume the least amount of electrical power?
    a. An air conditioner.
    b. An electric range.
    c. A water heater.
    d. A television.

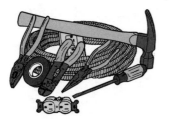

3
Electrical Circuit Components

e _____   Score _____

_____ Period _____   Instructor _____

ECTIVE: You will be able to describe the characteristics of various circuit components.

CTIONS: Carefully read Chapter 3 of the text. Then complete the following questions and problems.

For all practical purposes, only _____ and _____ are used in residential and commercial wiring.

Which of the following is the most common of conductor materials for electrical wiring because of its low cost and desirable characteristics such as strength and resistance to oxidation?

a. Aluminum.

b. Copper.

c. Rubber.

d. Plastic.

Which of the following conductor materials is subject to problems as a result of oxidation and expansion?

a. Tin.

b. Rubber.

c. Aluminum.

d. Copper.

List three factors that determine the size and type of wires to be used for electrical wiring.

_____

_____

_____

Name and describe the item shown below.

_____

_____

_____

_____

WIRE

6. The _____ (larger, smaller) the wire number, the smaller its diameter is.

7. The standard for wire sizes is called the _____ _____ _____.

8. List and describe two reasons why wire size is important.

_____

_____

_____

_____

9. What is the equation used to compute voltage drop?

10. What would be the voltage drop of 50 ft. of AWG 12 copper wire carrying 15 A? (Show your work below.)

_____ voltage drop

11. The most common coverings for wire used in electrical wiring are _____ or

_____.

12. Complete the chart below by filling in the type of covering indicated by the letter designations.

INSULATED WIRE COVERINGS

Covering Types	Letter Designation
1.	RH, RHH, RHW, RUH, RUW
2.	T, TW, THW, TBS, THHN
3.	TA
4.	SA
5.	A
6.	V
7.	AVA, AVL, AVB

13. What wire type would you use for dry/wet conditions at a maximum temperature of 167 °F (75 °C)? (You may wish to refer to the chart in Fig. 3-8 of the text.) _____

14. Label the various types of information marked on a cable covering.

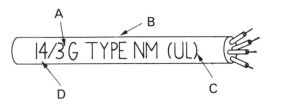

A.
B.
C.
D.

A. _____

B. _____

C. _____

D. _____

Electrical Circuit Components

15. Label the following wires as neutral or grounded, grounding conductor, or hot.

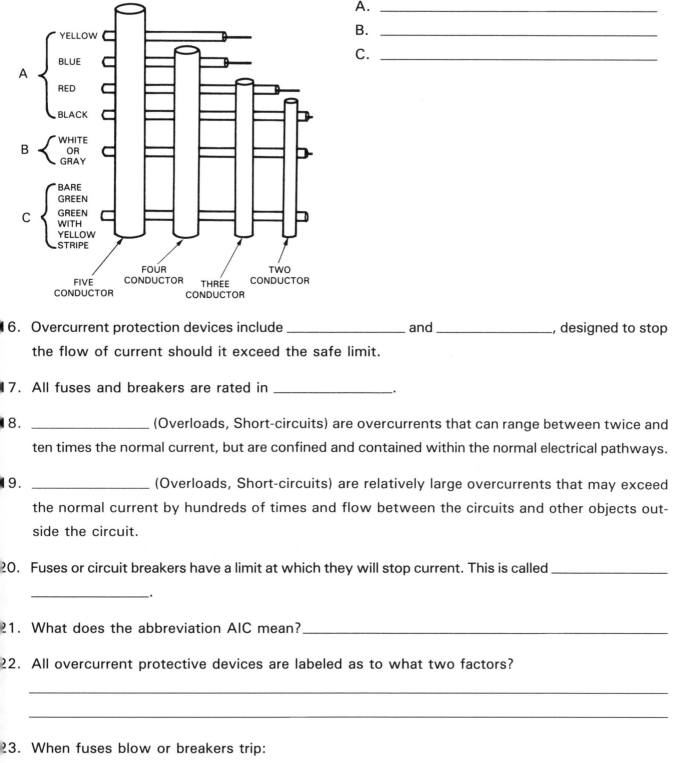

A. _____

B. _____

C. _____

16. Overcurrent protection devices include _____ and _____, designed to stop the flow of current should it exceed the safe limit.

17. All fuses and breakers are rated in _____.

18. _____ (Overloads, Short-circuits) are overcurrents that can range between twice and ten times the normal current, but are confined and contained within the normal electrical pathways.

19. _____ (Overloads, Short-circuits) are relatively large overcurrents that may exceed the normal current by hundreds of times and flow between the circuits and other objects outside the circuit.

20. Fuses or circuit breakers have a limit at which they will stop current. This is called _____ _____.

21. What does the abbreviation AIC mean?_____

22. All overcurrent protective devices are labeled as to what two factors?

_____

_____

23. When fuses blow or breakers trip:

a. The fuses or breakers should be replaced with protective devices or a higher amperage rating than the one intended for the circuit.

b. Ignore it.

c. The cause should be checked out before the fuses are replaced or the breakers are reset.

d. None of the above.

24. _____ are used to operate all, or sections (branches), of any circuit.

25. All switches are placed in:
    a. "Hot" conductors.
    b. Neutral conductors.
    c. Both "hot" or neutral conductors.
    d. None of the above.

26. List three different types of switches and their purposes.

    _____

    _____

    _____

    _____

    _____

    _____

27. What is the purpose of the electrical circuit?

    _____

28. Which of the following loads offer less resistance?
    a. A toaster.
    b. An ordinary light bulb.
    c. A waffle iron.
    d. An electric clock.

29. List four components of an electrical circuit in which resistance is present.

    _____   _____

    _____   _____

30. _____ is the amount of resistance which will permit the flow of 1 ampere of current when 1 volt of electrical force or pressure is applied.

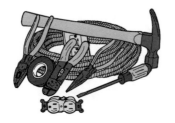

4
Tools for the Electrician

e _____   Score _____

_____ Period _____   Instructor _____

ECTIVE: You will be able to identify and properly use tools used for installing electrical wiring.

CTIONS: Carefully read Chapter 4 of the text. Then complete the following questions and problems.

tify each of the following tools and, in your own words, briefly describe their purpose(s).

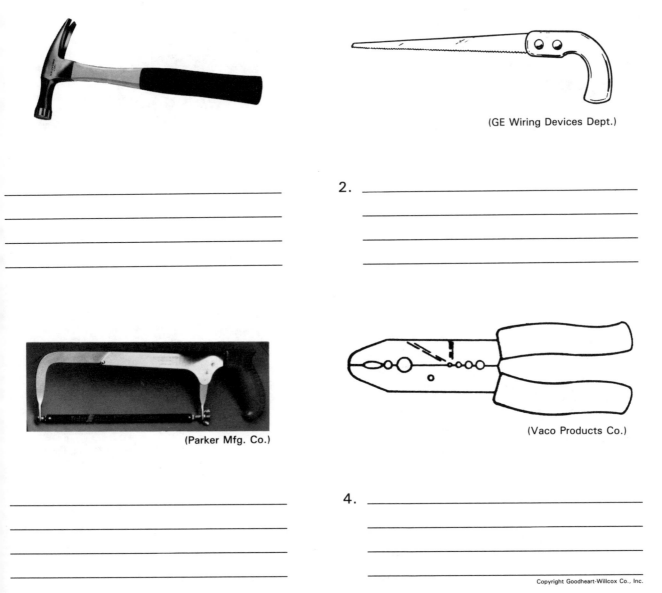

(GE Wiring Devices Dept.)

_____

2. _____

_____

_____

_____

(Parker Mfg. Co.)

(Vaco Products Co.)

_____

4. _____

_____

_____

_____

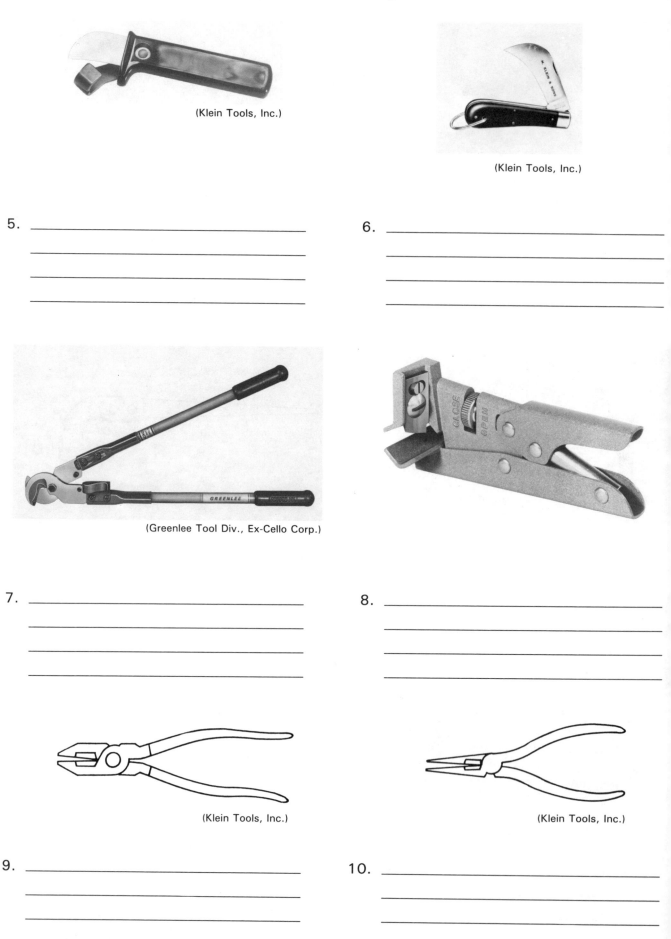

(Klein Tools, Inc.)

(Klein Tools, Inc.)

5. _____
   _____
   _____
   _____

6. _____
   _____
   _____
   _____

(Greenlee Tool Div., Ex-Cello Corp.)

7. _____
   _____
   _____
   _____

8. _____
   _____
   _____
   _____

(Klein Tools, Inc.)

(Klein Tools, Inc.)

9. _____
   _____
   _____
   _____

10. _____
    _____
    _____
    _____

Tools for the Electrician

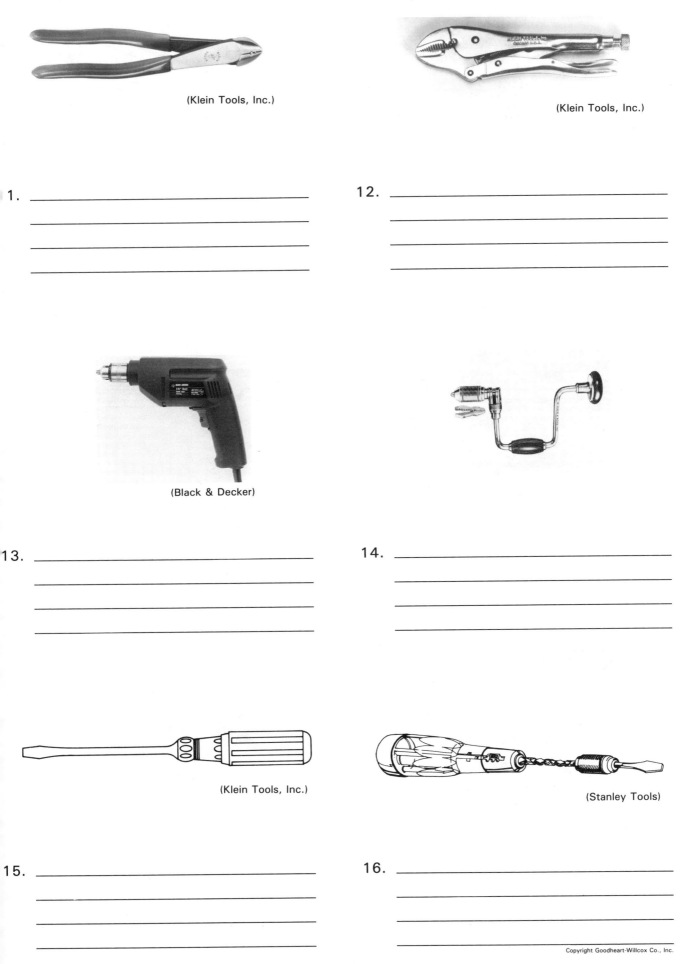

(Klein Tools, Inc.)

(Klein Tools, Inc.)

1. _____
   _____
   _____
   _____

12. _____
    _____
    _____
    _____

(Black & Decker)

13. _____
    _____
    _____
    _____

14. _____
    _____
    _____
    _____

(Klein Tools, Inc.)

(Stanley Tools)

15. _____
    _____
    _____
    _____

16. _____
    _____
    _____
    _____

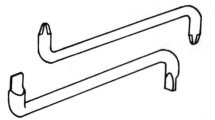

17. _____
_____
_____
_____

18. _____
_____
_____
_____

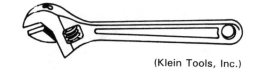

(Klein Tools, Inc.)

(Klein Tools, Inc.)

19. _____
_____
_____
_____

20. _____
_____
_____
_____

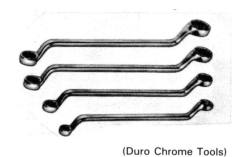

(Duro Chrome Tools)

21. _____
_____
_____
_____

22. _____
_____
_____
_____

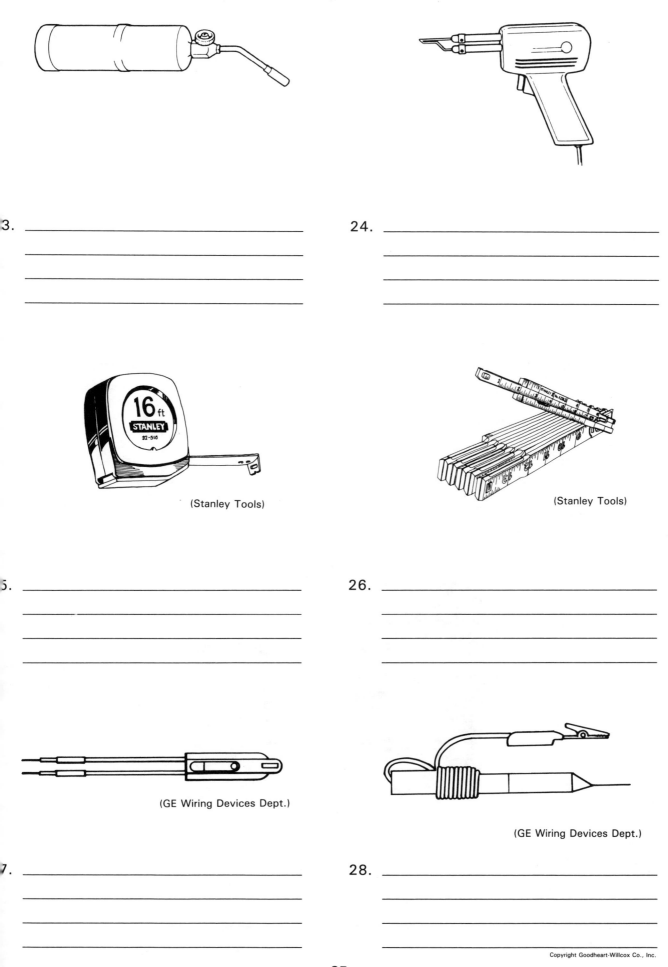

23. _____
_____
_____
_____

24. _____
_____
_____
_____

(Stanley Tools)

(Stanley Tools)

25. _____
_____
_____
_____

26. _____
_____
_____
_____

(GE Wiring Devices Dept.)

(GE Wiring Devices Dept.)

27. _____
_____
_____
_____

28. _____
_____
_____
_____

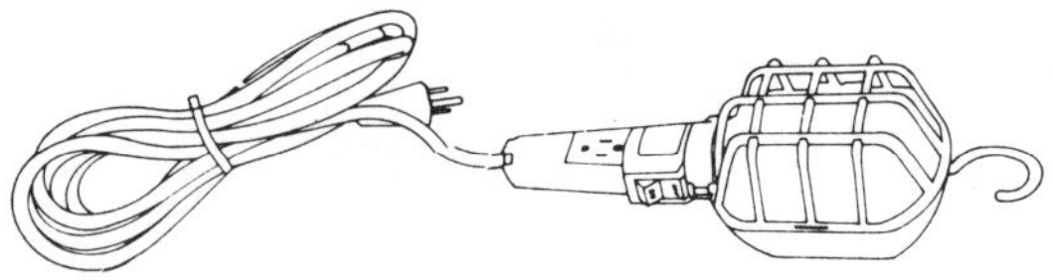

(GE Wiring Devices Dept.)

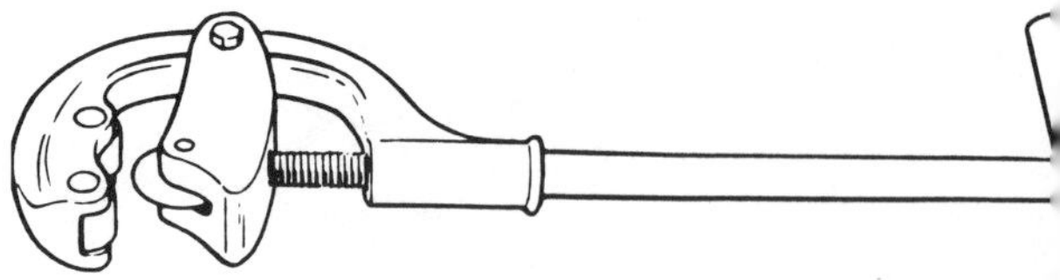

(GE Wiring Devices [

29. _____

30. _____

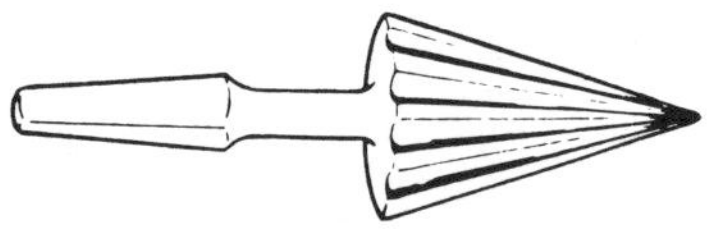

(GE Wiring Devices Dept.)

(GE Wiring Devices Dept.)

31. _____

32. _____

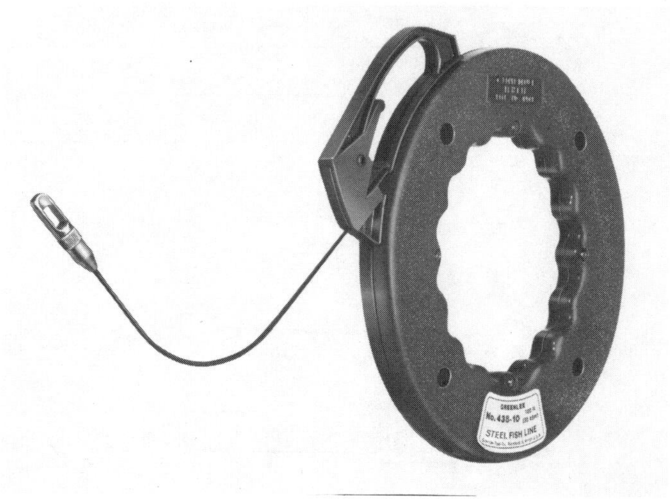

(Greenlee Tool Div., Ex-Cello Corp.)

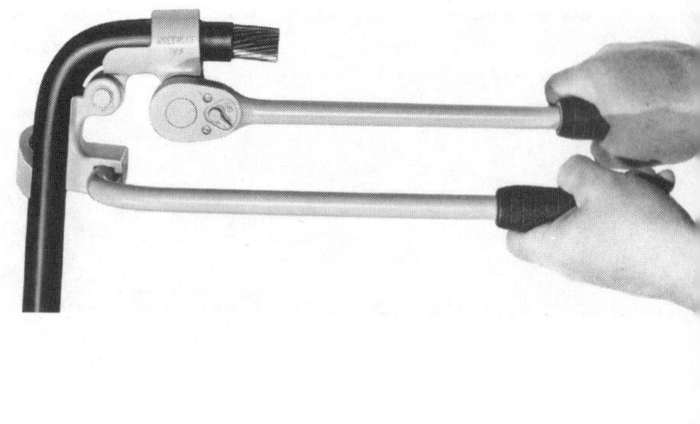

33. _____

34. _____

Tools for the Electrician

(Greenlee Tool Div., Ex-Cello Corp.)

(Appleton Electric Co.)

35. _____

_____

_____

_____

36. _____

_____

_____

_____

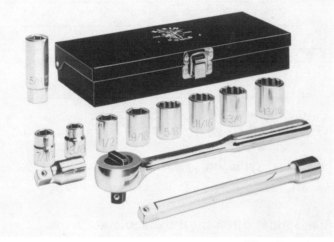

(Klein Tools, Inc.)

37. _____

_____

_____

_____

38. If any tool is damaged or worn, and reconditioning is impossible:

    a. Give it to someone else.

    b. The tool should be replaced.

    c. Use it carefully.

    d. None of the above.

39. List and briefly describe 10 tools that are essential to an electrician when wiring.

_____

_____

_____

_____

_____

_____

_____

_____

_____

_____

40. List four general rules regarding the proper use of striking tools.

_____

_____

_____

_____

_____

_____

_____

_____

41. The edges of cutting tools should always be kept _____.

42. For electrical work, use plier-type tools having properly _____ handles to help protect against electrical shock.

43. The fastening tool used most often by the electrician is the:
    a. Standard screwdriver.
    b. Phillips screwdriver.
    c. Open end wrench.
    d. Socket wrench.

44. When selecting a trouble light, be sure to select one with a _____ outlet.

45. Electrical contractors who hire electricians to work for them:
    a. Will provide the necessary tools for the worker.
    b. Will not expect the worker to have his or her own tools.
    c. Will expect the worker to have his or her own tools.
    d. None of the above.

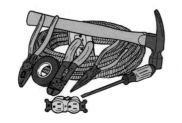

5
Safety and
Grounding Essentials

Score _____

_____
_____ Period _____   Instructor_____

CTIVE: You will be able to practice safety in electrical wiring. You will also be able to install a well-grounded system that follows *Code* and that will be serviceable and safe.

TIONS: Carefully read Chapter 5 of the text. Then complete the following questions and problems.

ist the 13 safety rules described in the text that should be learned and followed faithfully while istalling electrical wiring.

_____
_____
_____
_____
_____
_____
_____
_____
_____
_____
_____
_____
_____
_____
_____
_____
_____
_____

adders used by an electrician should be made of _____ or _____.

f a metal scaffolding is to be used, it should be _____.

When can temporary wiring be used?

a. On a construction site.

b. For emergency situations.

c. For Christmas lighting or carnival power.

d. All of the above.

ist two devices used to protect personnel on construction sites from electrical shock.

_____
_____

6. If the floor or clothing is wet or if a person is perspiring, electrical shock will be _____ (less, more) severe than if conditions are dry.

7. On which type of floor would the most severe shock result?
   a. Wood.
   b. Concrete.
   c. Metal.
   d. All would be equally severe.

8. Voltage does not have as much to do with severity of shock as the amount of current. (true or false)

9. Complete the following chart, describing the effects of electric current on the body.

AVERAGE EFFECTS OF ELECTRIC CURRENT ON BODY	
Amount of current (in amperes)	Effects on body
0.001 1 milliampere	
0.001 to 0.01 1 mA to 10 mA	
0.01 to 0.1 10 mA to 100 mA	
0.1 or more 100 mA or more	

10. If a shock victim is still in contact with the source of electrical current:
   a. Move the conductor or victim with your hands.
   b. Use a stick or insulated material to knock the conductor off of the victim.
   c. Avoid shutting off the power.
   d. Wait for help to arrive.

11. Define grounding.

   _____

   _____

12. What does grounding protect?

   _____

   _____

13. List three sources from which electrical systems could receive extremely high voltage.

   _____

   _____

   _____

14. List two kinds of grounds for electrical wiring.

   _____

5. _____ grounding is the intentional connection of one conductor of the electrical system to the earth.

6. Explain why grounding of electrical systems is required.

_____

_____

_____

_____

7. Label the various parts indicated on this schematic illustration of both system and equipment grounding.

A. _____

B. _____

C. _____

D. _____

E. _____

F. _____

G. _____

H. _____

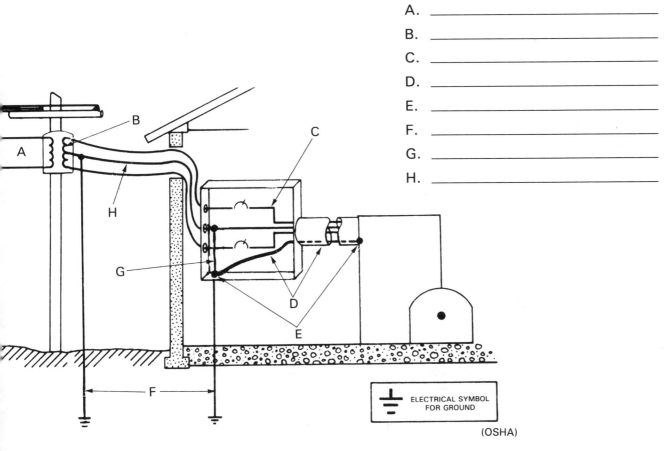

ELECTRICAL SYMBOL
FOR GROUND

(OSHA)

8. A ground rod must be driven to a depth of at least _____ ft. to assure a good connection of the neutral service conductor with the earth.

9. Grounding electrodes are connected to the neutral conductor by a stranded copper conductor no smaller than No. _____ AWG known as the grounding electrode conductor.

10. Another method of system grounding is to connect the neutral wire to the metal piping of the _____ supply system in the building.

11. Joining all metal parts of the wiring system—boxes, cabinets, enclosures, and conduit—to ensure having good, continuous metallic connections throughout the grounding system is called

_____.

22. Bonding is required at:

    a. All conduit connections of the electrical service equipment and points where a nonconduc ing substance is used that might impair continuity.

    b. All service equipment enclosures whether inside or outside the building.

    c. All metallic components of the electrical system, which are normally non-current carryin

    d. All of the above.

23. In theory, with the neutral wire grounded at the supplier's pole (transformer) or at the buildin what would happen if you touched the neutral wire?

    _____

    _____

24. _____ grounding is the method which bonds the grounding conductor to equipme enclosures and metallic noncurrent carrying equipment.

25. A _____ _____ occurs when conductor insulation fails or when a wi comes loose from its terminal point and contacts normally nonconducting metal parts.

26. Label the parts indicated on the service panel sketch below.

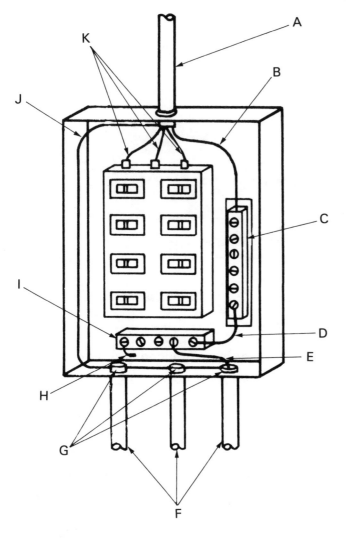

A. _____

B. _____

C. _____

D. _____

E. _____

F. _____

G. _____

H. _____

I. _____

J. _____

K. _____

Label the parts indicated on the subpanel sketch below.

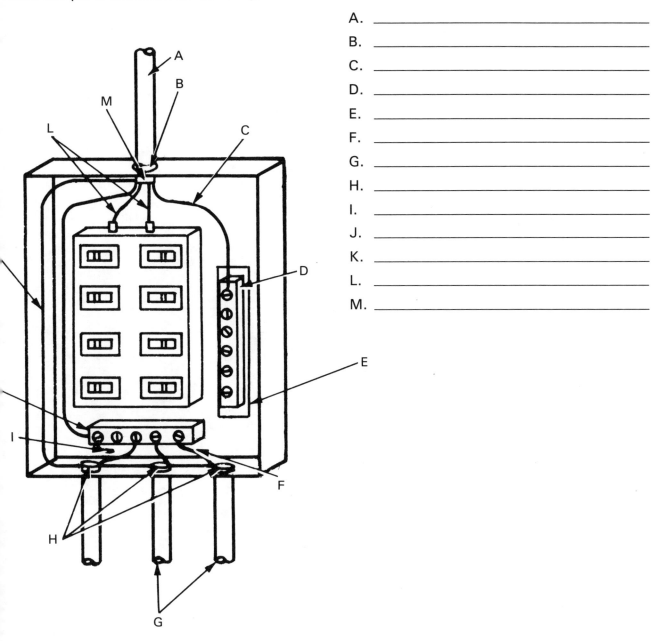

A. _____

B. _____

C. _____

D. _____

E. _____

F. _____

G. _____

H. _____

I. _____

J. _____

K. _____

L. _____

M. _____

What does the abbreviation GFCI mean?

_____

List three locations of a residence where the NEC requires GFCI protected circuits to be used.

_____

_____

_____

_____

_____

30. Name the device shown below and describe what it does.

_____

_____

_____

6
Wiring Systems

e _____   Score _____

_____ Period _____   Instructor_____

ECTIVE: You will be able to recognize various wiring systems. You will also be able to describe various wiring systems and determine where to use them.

CTIONS: Carefully read Chapter 6 of the text. Then complete the following questions and problems.

Conducting material for carrying electricity is commonly called _____.

List four items included in a wiring system.

_____

_____

_____

_____

Regardless of the system of wiring chosen or required, it is important to have a continuous _____ throughout every part of the system and every circuit.

Suppose you are working with a common 120 V ac circuit that has three wires. What is the purpose of each wire?

_____

_____

_____

List the eight wiring systems that are most often used today.

_____

_____

_____

_____

_____

_____

_____

_____

6. Identify the wiring systems shown below.

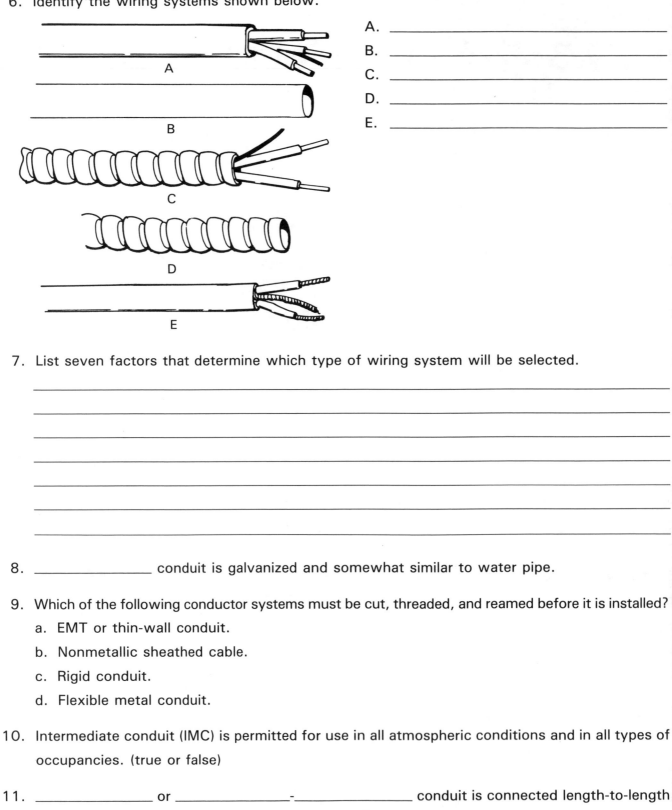

A. _____

B. _____

C. _____

D. _____

E. _____

7. List seven factors that determine which type of wiring system will be selected.

_____

_____

_____

_____

_____

_____

_____

_____

8. _____ conduit is galvanized and somewhat similar to water pipe.

9. Which of the following conductor systems must be cut, threaded, and reamed before it is installed?

    a. EMT or thin-wall conduit.

    b. Nonmetallic sheathed cable.

    c. Rigid conduit.

    d. Flexible metal conduit.

10. Intermediate conduit (IMC) is permitted for use in all atmospheric conditions and in all types of occupancies. (true or false)

11. _____ or _____-_____ conduit is connected length-to-length or to electrical boxes with suitable pressure couplings and connectors.

12. According to the *National Electrical Code,* when three or more conductors are used in conduit, the fill, for the most part, must not exceed _____ percent of the conduit's cross-sectional area.

13. When trying to find the correct conduit size, you should refer to tables in the _____ _____ _____.

14. Wires should be pulled through conduit only after all conduit is:
    a. Connected to all outlet boxes.
    b. Thoroughly tightened.
    c. Checked for burrs which could damage the insulation on the wire.
    d. All of the above.

15. Flexible metal conduit (Greenfield) is used outdoors in all kinds of weather conditions. (true or false)

16. Explain why Greenfield is frequently used with EMT or rigid conduit.

    _____

    _____

17. Because of its extreme flexibility, _____-_____ _____ _____ conduit can be used to connect machinery which is portable and/or which vibrates during normal operation.

18. Armored cable (BX) must only be used in dry locations since its cover is not weatherproof. (true or false)

19. Today, the most widely used conductor system used in residences is _____ _____ cable.

20. This type of nonmetallic sheathed cable may be used only in dry locations:
    a. NM.
    b. NMC.
    c. UF.
    d. USE.

21. This type of nonmetallic sheathed cable is extremely tough and durable, and used for direct burial.
    a. NM.
    b. NMC.
    c. UF.
    d. USE.

22. Rigid nonmetallic conduit:
    a. Can be used to support fixtures.
    b. Can be used in hazardous locations where physical damage is likely.
    c. Is corrosion-proof.
    d. All of the above.

23. PVC or nonmetallic conduit can be bent using special bending boxes which consist of _____ _____ to warm the conduit.

24. List four items that are needed in order to install PVC.

    _____

    _____

    _____

    _____

25. List five restrictions and limitations to consider in the use of electrical nonmetallic tubing (ENT).

    _____

    _____

    _____

    _____

    _____

26. What factor may limit the use of PVC in certain locations?

    _____

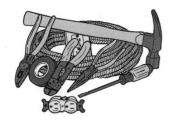

7
Boxes, Fittings, and Covers

e _____   Score _____

e _____ Period _____   Instructor _____

ECTIVE: You will be able to identify and determine where to use common boxes, fittings, and covers.

CTIONS: Carefully read Chapter 7 of the text. Then complete the following questions and problems.

_____ and _____ are used to house and protect electrical conductors and electrical devices.

The NEC requires that which of the following be housed inside approved enclosures?

a. Joints.

b. Connections.

c. Splices.

d. All of the above.

Name three materials from which electrical boxes may be made.

_____

_____

_____

Identify the four common box shapes used for electrical wiring boxes.

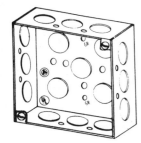

B

D

(Appleton Electric Co.)

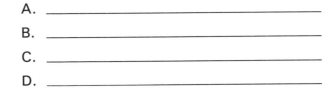

A. _____

B. _____

C. _____

D. _____

5. Which of the following shapes for electrical boxes is used only for fixtures such as ceiling lights?
   a. Square.
   b. Round.
   c. Octagonal.
   d. Rectangular.

6. Which of the following shapes for electrical boxes is preferred for wall receptacles and switches?
   a. Square.
   b. Round.
   c. Octagonal.
   d. Rectangular.

7. Metal boxes usually have a _____ finish.

8. Any type or shape of box is suitable as a pull box. (true or false)

9. _____ boxes are seamless and have rounded corners to prevent injuries.

10. _____ may be removed only to provide an opening for a cable, conduit, or fitting. No other openings are permitted in electrical boxes.

11. Boxes with removable sides are the only ones which can be ganged. (true or false)

12. Boxes must be securely fastened to a _____ member of the wall, ceiling, or floor of the dwelling.

13. _____ and _____ boxes are rugged, lightweight, and resist corrosion.

14. _____ are parts of a wiring system which are designed to interconnect conduit, conductors, or boxes.

15. Name 10 types of fittings used in electrical wiring.

   _____
   _____
   _____
   _____
   _____
   _____
   _____
   _____
   _____
   _____

6. List three purposes of box extension rings.

_____

_____

_____

7. List two fittings which secure conductors to an electrical box.

_____

_____

8. _____ devices are used to prevent strain on wire connections made to outlets or switches.

9. Grounding is required only on metal boxes. (true or false)

0. _____ _____ is a spring or tension device that holds the ground wire in tight electrical contact with the box when it is hooked over the edge of the box.

1. NEC requires a grounding _____ wherever conduit enters a nonthreaded opening on a box.

2. Identify the switch and box covers shown below.

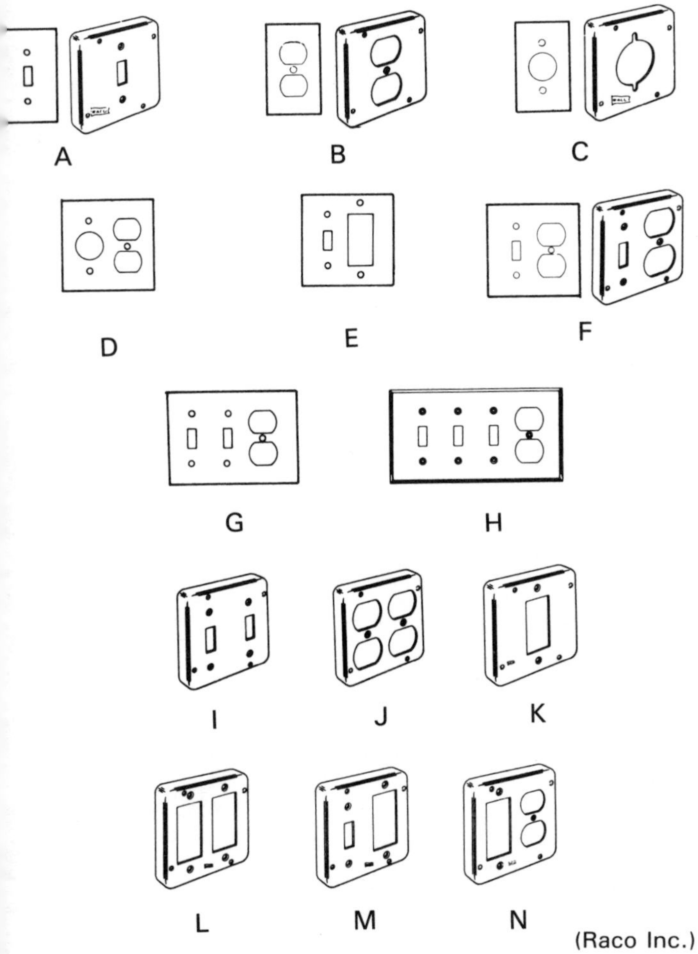

A
B
C
D
E
F
G
H
I
J
K
L
M
N

A. _____

B. _____

C. _____

D. _____

E. _____

F. _____

G. _____

H. _____

I. _____

J. _____

K. _____

L. _____

M. _____

N. _____

(Raco Inc.)

23. _____ _____ refers to the number of conductors that the *Code* will allow in certain sizes of boxes.

24. Plaster rings are special adaptors that can be used on either ceiling or wall boxes. (true or false)

25. List three ways of increasing the capacity of a box.

    _____

    _____

    _____

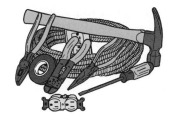

8
Installing Boxes and Conductors

Score _____

_____

_____ Period _____

Instructor _____

OBJECTIVE: You will be able to rough-in and install boxes, conduit, cable, and conductors.

DIRECTIONS: Carefully read Chapter 8 of the text. Then complete the following questions and problems.

Boxes, conduit, cable, and conductors are installed in new construction while the walls are still open. (true or false)

Before beginning a _____ - _____, an electrician will prepare a rough sketch of a room to indicate where and what types of electrical components are required.

Switches and other devices intended for standard elbow height are usually marked at _____ inches from the rough floor.

Outlet boxes are customarily attached to the stud _____ inches off the floor.

Most electricians run cable or conduit before boxes are installed. (true or false)

Identify the various methods used for attaching boxes as shown below.

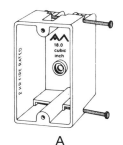

A

A. _____

B. _____

C. _____

D. _____

E. _____

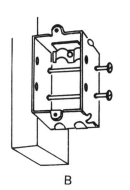

B

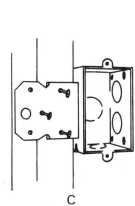

C

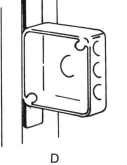

D

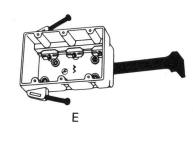

E

7. A box must be mounted so that its face extends beyond the stud and the thickness of the drywall of plaster surface. (true or false)

8. A _____ _____ is the section from the circuit breaker to the first device in the circuit.

9. Before _____ is cut and bent, you will need to plan and measure the best route from box to box and then to the panel.

10. Notching for conduit is preferred since it does not weaken the frame as much as boring does. (true or false)

11. Name two tools normally used to cut all types of conduit.

_____

_____

12. _____ bends are those made on the construction site.

13. A _____ is used primarily to bend large sizes of conduit while a _____ is used to bend smaller sizes of conduit.

14. NEC specifies the _____ of conduit bends.

15. Define stub height distance.

_____

_____

16. Sketch a back-to-back bend in the space provided.

17. Explain when an offset bend would be used.

_____

_____

Installing Boxes and Conductors

8. List four facets that must be considered when producing an offset bend.

_____

_____

_____

_____

9. What is the key to precise bending?

_____

0. Explain the difference between the saddle bend and the offset bend.

_____

_____

_____

1. When making the saddle bend, the center bend is always made first. (true or false)

2. When bending nonmetallic conduit by hand, use heat resistant _____.

3. When bending IMC or EMT, the number of bends between outlets or fittings must not be greater than the equivalent of four 90° bends or 360° total. (true or false)

4. In general, metal conduit must be supported within _____ ft. of every outlet box and at a minimum of every _____ ft. of run.

5. When installing flexible metal conduit, support must be provided at least every _____ ft. and within _____ in. of every outlet or fitting.

6. List two reasons why *Code* requires that conduit is securely fastened to electrical boxes.

_____

_____

7. A _____ _____ is used to pull wire through conduit.

8. Wires being pulled should be kept _____ (bent, straight).

9. _____ cable is quite flexible and can be pulled through bored holes with greater ease than conduit.

0. Armored cable should be measured off and cut _____ (before, after) it is pulled through the holes in the framing members.

1. List three methods of cutting armored cable.

_____

_____

_____

32. Identify the parts indicated in the drawing below that illustrate installing armored cable to a bo

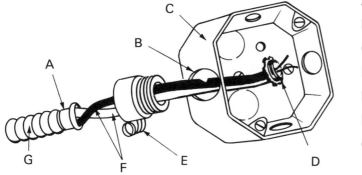

A. _____

B. _____

C. _____

D. _____

E. _____

F. _____

G. _____

33. The outer covering of nonmetallic sheathed cable is usually a plastic or thermoplastic materi
    (true or false)

34. Why are cable strippers used?

    _____

    _____

35. Ground wires should be attached:

    a. Together.

    b. To the box.

    c. To any device installed at the box.

    d. To all of the above.

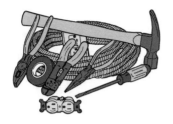

9
Device Wiring

Name _____ Score _____

_____ Period _____ Instructor _____

OBJECTIVE: You will be able to splice conductors to one another and connect them to switches, receptacles, and fixtures.

DIRECTIONS: Carefully read Chapter 9 of the text. Then complete the following questions and problems.

1. _____ refers to the installation of devices and fixtures.

2. If an electrical product carries the label of a testing agency, the *Code* states that you can use the product for any purpose you choose. (true or false)

3. Conductors and cables must be carefully routed and sufficiently supported to eliminate "bunching" or "twisting" of the cable and possible conductor damage. (true or false)

4. All splices and connections must be covered with an insulation _____ (less than or equal to) the conductor's original insulation.

5. *Code* requires that _____ switches for electrical machines and devices should be clearly marked to identify what they control.

6. In removing insulation from a conductor, you must be careful not to damage either the conductor _____ or the remaining insulation.

7. Which wire below has been properly stripped? _____

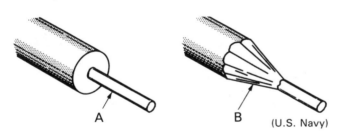

A   B   (U.S. Navy)

8. As a general rule, the amount of conductor that should be bared to allow enough bare wire to make a proper connection is:

a. 7/8 inch.

b. 1/8 inch.

c. 5 inches.

d. 1 foot.

9. The preferred way to remove insulation is with a _____ _____.

10. When attaching a conductor to a device terminal, the curved hook on the conductor must be connected:
    a. So that it has less than a two thirds wrap.
    b. Counterclockwise onto the terminal.
    c. Clockwise onto the terminal.
    d. So that it overlaps.

11. Conductors are spliced together by twisting the wires together in a _____ (clockwise counterclockwise) direction.

12. The steps for producing a Western Union splice are shown below. Place them in the correct order by numbering the steps from 1 to 5.

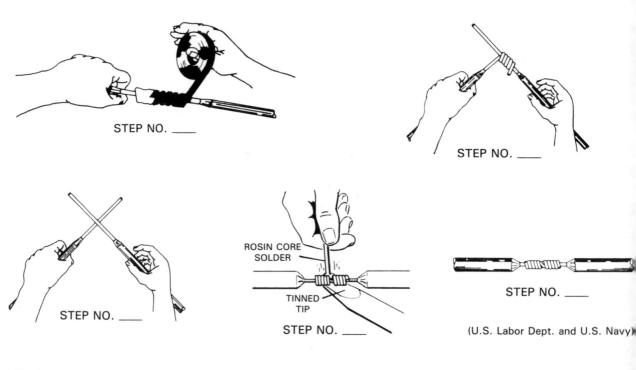

STEP NO. ____

STEP NO. ____

STEP NO. ____

ROSIN CORE SOLDER

TINNED TIP

STEP NO. ____

STEP NO. ____

(U.S. Labor Dept. and U.S. Navy)

13. The key to good _____ is to heat the joint so well that it melts the solder onto itself.

14. What can you use to join large-diameter conductors?

    _____

15. The function of a _____ is to control the flow of electricity to one or more electrical devices.

16. List the two common switch ratings.

    _____

    _____

17. A single-pole switch has four terminals. (true or false)

Device Wiring

Outlet _____ are used to transfer electrical energy from conductors to appliances.

Identify the parts indicated on the duplex receptacle below.

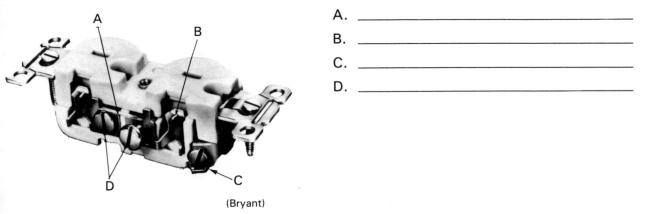

(Bryant)

A. _____

B. _____

C. _____

D. _____

Explain how to connect a grounding type receptacle to conductors.

_____

_____

_____

What method of connecting several wires together is shown in the diagram below?

_____

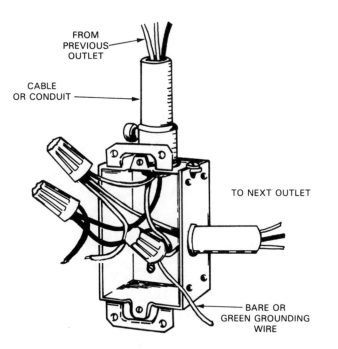

FROM PREVIOUS OUTLET

CABLE OR CONDUIT

TO NEXT OUTLET

BARE OR GREEN GROUNDING WIRE

A _____-_____ receptacle is one in which one of the outlets in a duplex receptacle is controlled by a switch while the other is always energized.

23. Identify the items indicated in the drawing below.

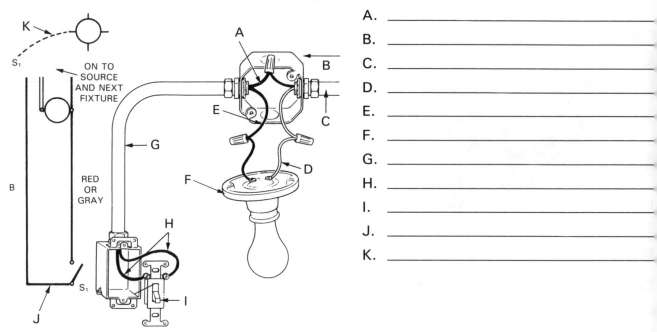

A. _____

B. _____

C. _____

D. _____

E. _____

F. _____

G. _____

H. _____

I. _____

J. _____

K. _____

24. Fixtures may be mounted to boxes in a variety of ways depending on:

    a. Type and location of the fixture.

    b. Material from which wall or ceiling is made.

    c. Weight of the fixture.

    d. All of the above.

25. Heavier fixtures require substantial box _____ and _____.

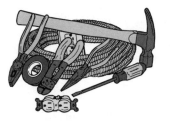

10
Planning Branch Circuits

e _____

e _____ Period _____

Score _____

Instructor _____

OBJECTIVE: You will be able to plan for the minimum requirements as well as for future needs for a well designed wiring system.

DIRECTIONS: Carefully read Chapter 10 of the text. Then complete the following questions and problems.

List seven problems that can result from obsolete wiring.

_____

_____

_____

_____

_____

_____

_____

_____

Explain why it would be impractical to place all electrical loads in a building on a single circuit.

_____

_____

_____

A _____ _____ is a separate electrical path, independent of other electrical paths in the building.

Name the four main categories of branch circuits.

_____

_____

_____

_____

What two factors must be considered when planning branch circuits?

_____

_____

The *Code* allows _____ VA per square foot of floor area.

7. The loads for lighting outlets and receptacle outlets can vary considerably. (true or false)

8. In determining electrical loads, _____ W (volt-amperes) should be allowed per grouped duplex receptacle outlets.

9. List seven factors that should be considered when designing a branch circuit.

   _____

   _____

   _____

   _____

   _____

   _____

   _____

10. It would be unwise to use wire smaller than No. _____ AWG in homes that are built today.

11. Lights should be controlled by:
    a. A switch or switches near the entrance(s) to the room.
    b. A pull chain on the fixture itself.
    c. A pushbutton on the fixture itself.
    d. All of the above.

12. A light switch should be located on the latch side of a doorway. (true or false)

13. Wall outlets must be located so that no point along any wall will be more than _____ ft (_____ mm) from a receptacle.

14. List nine places where receptacle outlets are required in a residence. Also indicate where GFCI receptacles should be used.

    _____

    _____

    _____

    _____

    _____

    _____

    _____

    _____

    _____

15. Receptacle outlets should be placed _____ to _____ inches above the floor line and as close to the ends of large wall spaces as possible.

16. Every room should have at least one lighting outlet. (true or false)

7. List nine appliances that require a separate circuit.

_____
_____
_____
_____
_____
_____
_____
_____
_____
_____

8. Determine the number of general lighting circuits needed for a dwelling with an occupied area of 1400 sq. ft. The ampacity of the circuits may be either 15 or 20 amperes. (Show your work in the space provided).

    1. Find out how many watts the *Code* requires of the occupied space.

        _____ VA

    2. Find the total amperage needed if the supplied voltage is 120 V.

        _____ Amperes

    3. Now, divide by the amperage of the circuits to find the number of circuits needed.

        _____ 15 ampere circuits minimum

        or

        _____ 20 ampere circuits minimum

19. When planning the hookup of branch circuits to the service panel, it is important to keep _____ in the load between the two hot wires in a three-wire system.

20. List five steps in determining how many branch circuits will be necessary.

_____

_____

_____

_____

_____

_____

Additional activity:

Using the information provided in this chapter, design an electrical plan for the single-family dwelling shown below. Draw in switching arrangement and general and special branch circuits. Have your instructor evaluate your electrical plan. (You may want to refer to Figure 11-4 of the text to see how to draw common electrical symbols.)

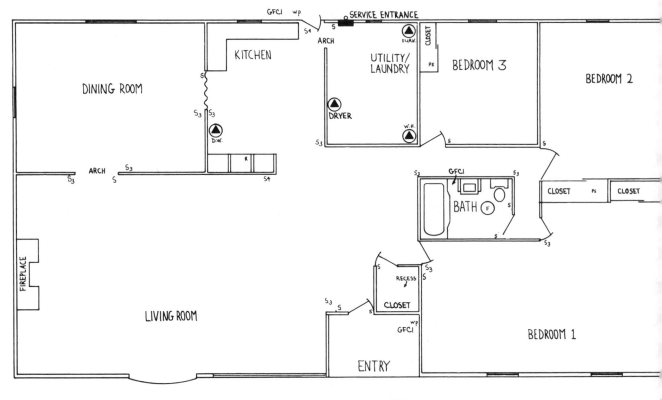

HOUSE AREA = 57' x 30' =1710 SQ. FT.

POWER EQUIPMENT = 1710 x 3 WATTS	=	5130 VA
TWO APPLIANCE CIRCUITS AT 1500 VA	=	3000 VA
ONE LAUNDRY CIRCUIT AT 1500 VA	=	1500 VA
SUBTOTAL SM. APPLIANCE AND LTG	=	9630 VA
WATER HEATER 7.5 kW	=	7500
DRYER 5.0 kW	=	5000
DISHWASHER 1.5 kW	=	1500
RANGE	=	17,000
FURNACE (omit air cond. NEC 220-30[c])	=	10,000 VA

FIRST 10,000 AT 100%, REMAINDER AT 40% = 26,252 (26.25 kW)
26,252 ÷ 240 V = 109 A = SERVICE MINIMUM

(A) NUMBER OF LIGHTING GENERAL PURPOSE CIRCUITS = 5130 ÷ 120 VOLTS = 42.75 AMPERES OR MINIMUM 3, 15 AMPERE CIRCUITS

(B) AND NUMBER OF SMALL APPLIANCE CIRCUITS = 2 at 20 AMP EACH, PLUS 1 LAUNDRY 20 AMPERE CIRCUIT

(C) PLUS SPECIAL PURPOSE CIRCUITS = 5
TOTAL BRANCH CIRCUITS (A)(B)(C) = 11 MINIMUM

RECOMMENDED SERVICE
120/240 VOLT, 3 WIRE
NO. 2 THW TYPE WIRE
COPPER CONDUCTORS

NOTE ALL CALCULATIONS ARE FOR THE MINIMUM REQUIREMENT

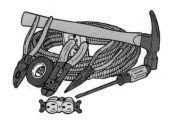

11
Reading Prints and Wiring Circuits

_____   Score _____

_____ Period _____   Instructor_____

CTIVE: You will be able to read prints (blueprints). You will also be able to identify various symbols used on an electrical plan and wire circuits.

CTIONS: Carefully read Chapter 11 of the text. Then complete the following questions and problems.

ELECTRICAL SYMBOLS

American National Standards Institute (ANSI) symbols should be understood and memorized ery electrician. Commonly used symbols that an electrician will need to know are given below. fy each of the following symbols.

RAL OUTLETS

Ⓞ   ─Ⓞ   _____

Ⓑ   ─Ⓑ   _____

Ⓓ   WALL   _____

Ⓔ   ─Ⓔ   _____

Ⓕ   ─Ⓕ   _____

Ⓙ   ─Ⓙ   _____

Ⓛ   ─Ⓛ   _____

ⓁPS   ─ⓁPS   _____

Ⓢ   ─Ⓢ   _____

Ⓥ   ─Ⓥ   _____

Ⓧ   ─Ⓧ   _____

Ⓒ   ─Ⓒ   _____

⌥   _____

CONVENIENCE RECEPTACLES

14. _____

15. _____

16. _____

17. _____

18. _____

19. _____

20. _____

21. _____

22. _____

23. _____

SWITCH OUTLETS

24. S _____

25. S_2 _____

26. S_3 _____

27. S_4 _____

28. S_D _____

29. S_E _____

30. S_K _____

31. S_P _____

32. S_{CB} _____

33. S_{WCB} _____

34. S_{MC} _____

35. S_{RC} _____

36. S_{WP} _____

37. S_F _____

38. S_{WF} _____

PANELS, CIRCUITS, AND MISCELLANEOUS

39. _____

40. _____

41. _____

42. _____

43. _____

44. _____

45. _____

46. _____

47. _____

48. _____

49. _____

50. _____

51. _____

52. _____

AUXILIARY SYSTEMS

53. _____

54. _____

55. _____

56. _____

57. _____

58. _____

59. _____

60. _____

61. _____

62. <u>F</u>O _____

63. <u>F</u> _____

64. <u>X</u> _____

65. <u>FA</u> _____

66. <u>FS</u> _____

67. <u>W</u> _____

68. <u>W</u> _____

69. <u>H</u> _____

70. <u>N</u> _____

71. <u>M</u> _____

72. <u>R</u> _____

73. <u>SC</u> _____

74. ☐ _____

75. ⑊⑊⑊ _____

76. Name three types of information that electrical symbols can give you.

_____

_____

_____

77. Explain why on an electrical print, the dashed line running from a switch symbol (S) to the outlet symbol is always curved.

_____

_____

78. A special outlet symbol carrying a lower case letter alongside must be listed in the key of symbols on each drawing and if necessary, further described in the specifications. (true or false)

79. A _____ or _____ layout is a print of the electrical plan which explains where cable is to run and how many conductors it must have.

80. List the three primary reasons for polarity wiring.

_____

_____

_____

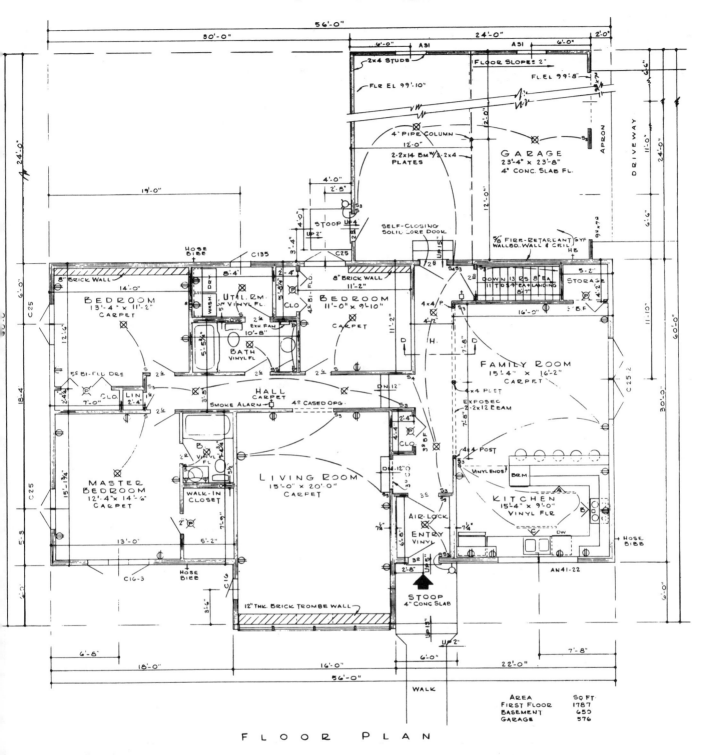

F L O O R   P L A N

AREA	SQ FT
FIRST FLOOR	1787
BASEMENT	655
GARAGE	576

31. In the floor plan above, how many duplex convenience receptacles are indicated for the living room?

_____

32. In the floor plan above, what type of switches are indicated for the bedrooms?

_____

33. In the drawing above, which rooms have three-way switches?

_____

_____

84. In the chart below, identify the name of the system and sketch each one as a pictorial.

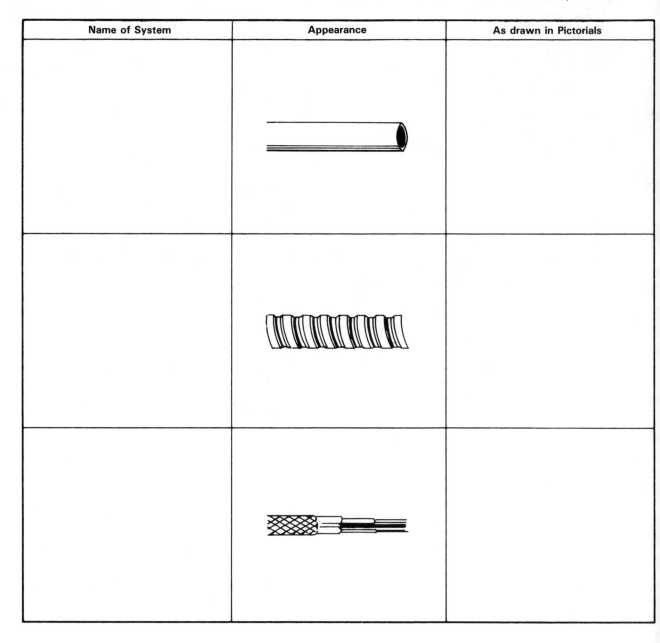

Name of System	Appearance	As drawn in Pictorials

85. In the spaces below, sketch the print and schematic of the pictorial given.

Pictorial	Print	Schematic

6. Duplex receptacles with provision for grounding can be continuously live or switched. (true or false)

7. _____-_____ receptacles are used mostly in rooms which have no ceiling outlets. Thus, lighting devices such as lamps can be switch controlled.

8. _____-_____switches control lights from two distant locations (for example, in rooms having two entrances).

9. A _____ switch is wired so that it glows when the fixture is on.

0. To protect other receptacles that follow a GFCI on the same circuit, the GFCI must be placed in the outlet that is electrically closest to the circuit breaker panel (power source). (true or false)

1. Draw the print plan of the pictorial shown below.

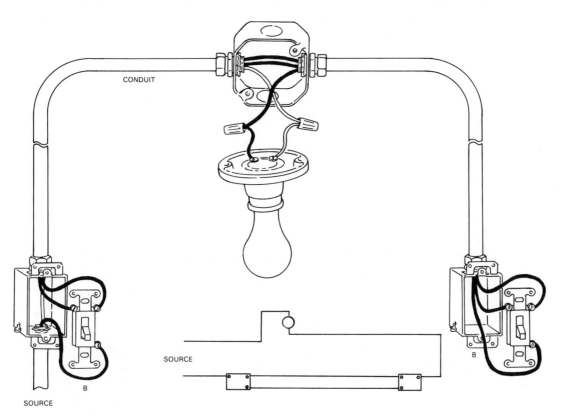

CONDUIT

SOURCE

B

SOURCE

B

Print plan:

92. Draw the schematic of the pictorial shown below.

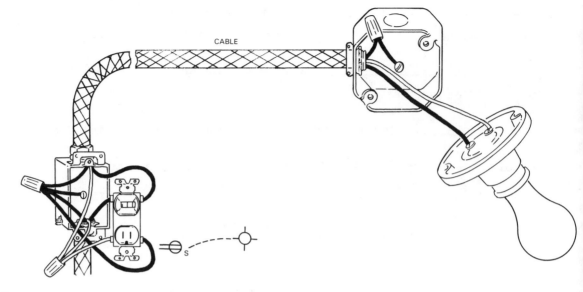

Schematic:

93. Using colored pencils, draw in the wires using the correct colors on the pictorial below. The pictorial shows a light controlled by a switch. The receptacle beyond the switch is live. The source is at the light.

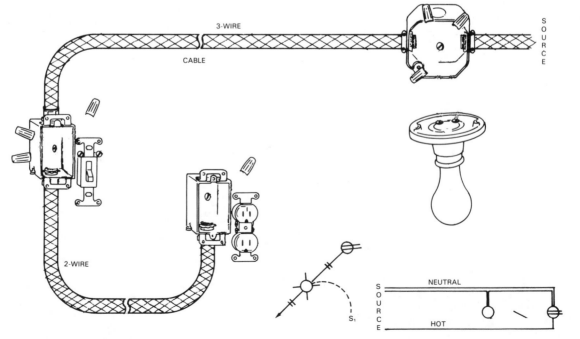

4. A light fixture controlled from four locations using two three-way and two four-way switches is shown below. The feed is at the fixture. In the space provided, draw the cable layout for wiring this circuit.

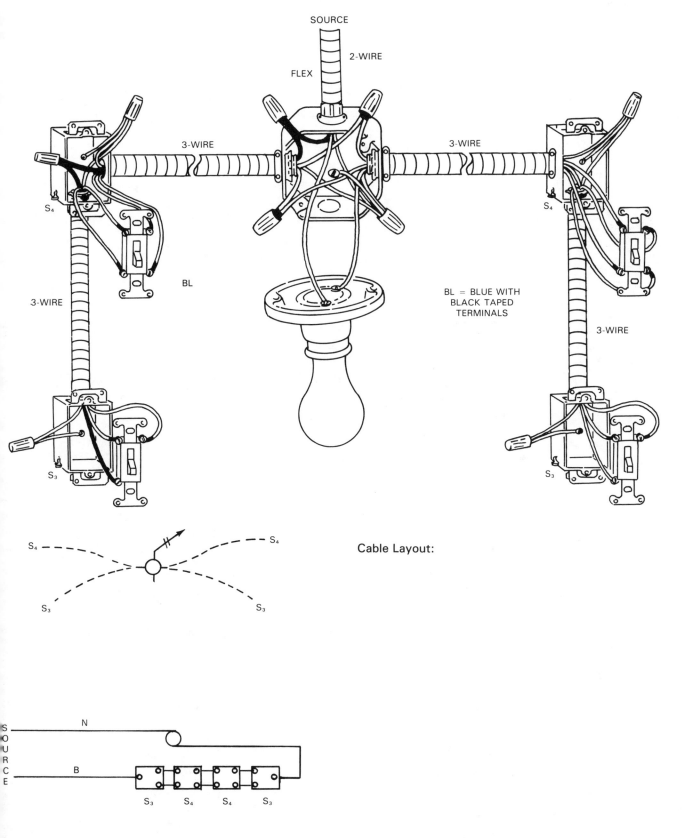

SOURCE

2-WIRE

FLEX

3-WIRE   3-WIRE

S_4 S_4

BL

3-WIRE

BL = BLUE WITH
BLACK TAPED
TERMINALS

3-WIRE

S_3 S_3

S_4 - - - - -   - - - S_4

S_3 - - - -   - - - S_3

Cable Layout:

SOURCE

N

B

S_3 S_4 S_4 S_3

95. Using colored pencils, draw in the wires using the correct colors in the pictorial below. The pictorial shows one light controlled by a single-pole switch. One circuit is the source, the single pole and one three-way switch are housed together.

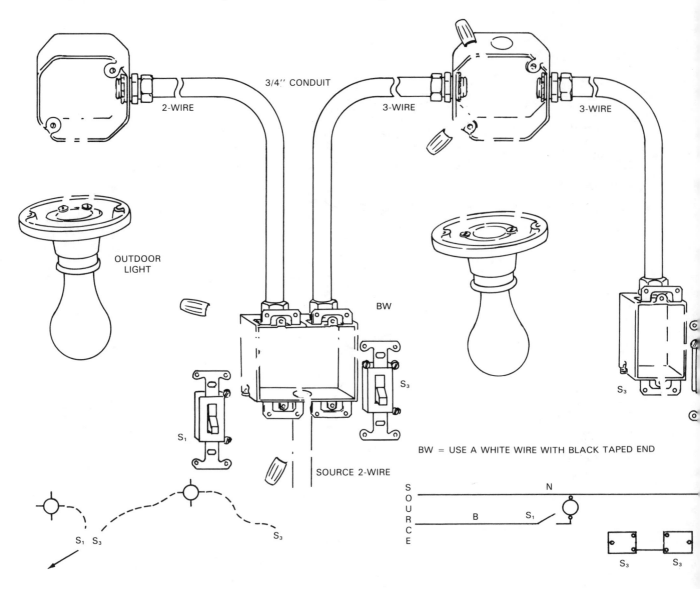

3/4'' CONDUIT

2-WIRE

3-WIRE

3-WIRE

OUTDOOR LIGHT

BW

BW = USE A WHITE WIRE WITH BLACK TAPED END

SOURCE 2-WIRE

96. Identify the feed-through outlet and the termination outlet by labeling them below.

A. _____

B. _____

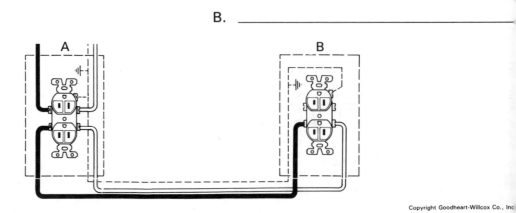

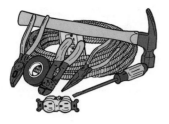

12
The Service Entrance

e _____   Score _____

_____ Period _____   Instructor _____

ECTIVE: You will be able to describe the components of the service entrance.

CTIONS: Carefully read Chapter 12 of the text. Then complete the following questions and problems.

All electrical energy supplied to power-consuming devices in a building must first pass through the _____ entrance equipment.

Power is carried from the pole transformer on three wires of which:

a. One is hot and two are grounded neutral wires.

b. Two are hot and one is a grounded neutral wire.

c. All are hot wires.

d. All are grounded neutral wires.

Service wires brought in to a building overhead from a utility pole are called the service _____.

Service wires routed underground from either a pole or transformer pad are called the service _____.

List eight guidelines an electrician should follow when choosing the location of a service entrance.

_____

_____

_____

_____

_____

_____

_____

_____

_____

For homes with electrical heating systems or where future expansion is likely, a _____ A or _____A service entrance load is best.

Most service drops in new homes are made with _____ cable.

8. Identify the items indicated on the service entrance diagram below.

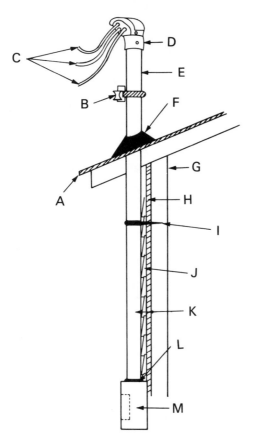

A. _____

B. _____

C. _____

D. _____

E. _____

F. _____

G. _____

H. _____

I. _____

J. _____

K. _____

L. _____

M. _____

9. A _____ _____ is a fitting installed at the top of the conduit or service entrance cable to prevent water from entering and shorting out the conductors.

10. The watt-hour _____ keeps records of electrical power consumed.

11. A weatherproof or watertight connector must always be used when the service entrance cable enters the meter enclosure from the top or side. (true or false)

12. An _____ is used as a means of attaching the service drop wires to the service entrance conductors.

13. A split-bolt _____ - _____ connector assures good electrical connection between large connectors.

14. List two types of equipment that will disconnect all wiring from the power source.

_____

_____

15. If a main disconnect is placed ahead of a lighting panel, the supply conductors must have an ampacity less than that of the main breaker or service switch. (true or false)

16. Generally, the main disconnect is part of the service panel, consisting of a main breaker or pair of fuses. (true or false)

7. Identify the items indicated on the service entrance layout shown below.

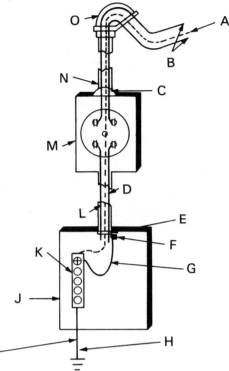

A. _____

B. _____

C. _____

D. _____

E. _____

F. _____

G. _____

H. _____

I. _____

J. _____

K. _____

L. _____

M. _____

N. _____

O. _____

8. List four characteristics of a grounding electrode conductor.

_____

_____

_____

_____

9. On the figures below, indicate the above-roof clearance for overhead conductors.

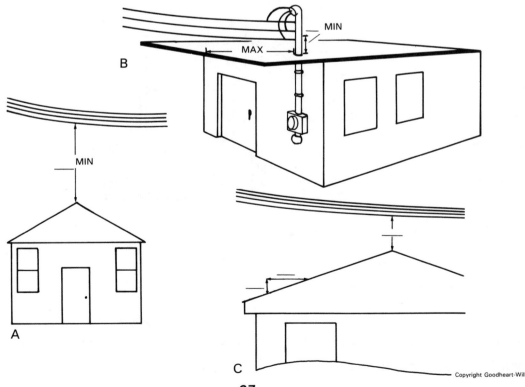

20. On the figure below, indicate the above-grade minimum clearances.

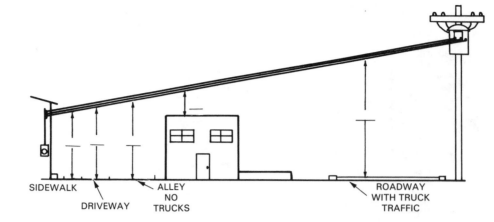

SIDEWALK   ALLEY NO TRUCKS   ROADWAY WITH TRUCK TRAFFIC

DRIVEWAY

21. On the figures below, indicate the clearances around building platforms and openings.

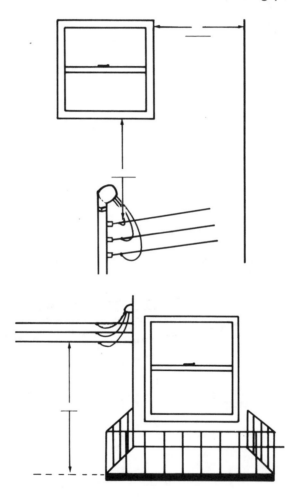

22. What is the purpose of circuit breakers and fuses?

_____

_____

23. A _____ _____ is a device designed to open and close a circuit by nonautomatic means and to open the circuit automatically on a predetermined overcurrent without injury to itself when properly applied within its rating.

The Service Entrance

24. Essentially, fuses perform the same function as breakers. (true or false)

25. What type of service rating is most often installed for one and two-family dwellings?

_____

26. The term _____-_____ indicates three separately derived ac currents which are "out of step" from each other by 120 electrical degrees.

27. The supply or source for a service entrance begins at the _____.

28. Transformers which increase the primary voltage are called step-down transformers. (true or false)

29. A chain of transformers links the generating plant with the consumer. (true or false)

30. Four-wire, three-phase systems are often used in structures such as:
    a. Factories.
    b. Hotels.
    c. Apartment complexes.
    d. All of the above.

31. The _____ (first or final) step-down occurs at the local pole transformer near the structure it serves.

32. Identify the items indicated on the diagram below.

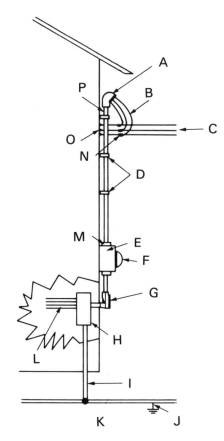

A. _____
B. _____
C. _____
D. _____
E. _____
F. _____
G. _____
H. _____
I. _____
J. _____
K. _____
L. _____
M. _____
N. _____
O. _____
P. _____

33. A more versatile alternative to single-phase wiring is _____-_____ wiring

34. _____-connected three-phase, four-wire service supplies 120 V single-phase, an
    240 V three-phase circuits.

35. In Delta four-wire systems, the ''wild'' phase C must always be identified at all terminations an
    accessible points. This is usually indicated by the color:
    a. Green.
    b. Purple.
    c. Orange.
    d. White.

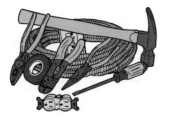

13
Appliance Wiring and Special Outlets

e _____   Score _____

_____ Period _____   Instructor _____

ECTIVE: You will be able to describe the wiring considerations and procedures for wiring appliances and special outlets.

ECTIONS: Carefully read Chapter 13 of the text. Then complete the following questions and problems.

List five large appliances that require special wiring considerations.

_____

_____

_____

_____

_____

_____ receptacle hookups are commonly used on dryers and ranges.

In general, appliances should be placed on a separate circuit if they are rated at:

a. 120 volt, 12 ampere plug.

b. 1/8 plus hp.

c. 240 volts or more.

d. All of the above.

Name the three methods by which heat travels.

_____

_____

_____

Heaters are almost always automatically controlled by _____ which are located close by the heating unit or built into it.

Latest NEC publications prefer giving heater sizes in _____-_____.

The most common residential and small commercial type heaters are _____ units.

8. Name six factors that can influence a choice of heater units.

   _____

   _____

   _____

   _____

   _____

   _____

9. The circuitry for a baseboard heater is shown below. Label the parts indicated.

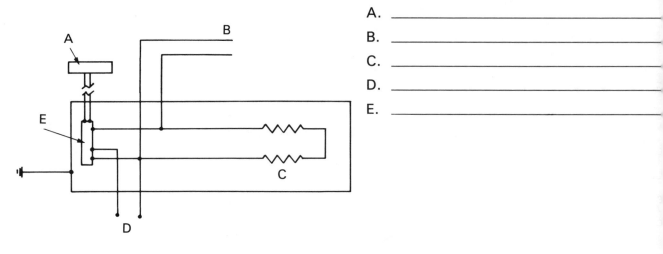

A. _____

B. _____

C. _____

D. _____

E. _____

10. Electric water heaters require a separate _____ volt circuit.

11. _____ _____ can often be connected to the same circuit as heating units since they would not be operating at the same time.

12. Garbage disposal units are most often controlled:

    a. By a plug and cord connection.

    b. From the panel circuit breaker.

    c. By an on-off switch.

    d. None of the above.

13. Dishwashers carry a rating within the range of 5-10 A at 120 V. (true or false)

14. Refrigerators and freezers should have overcurrent protection, fuses, or breakers rated at no more than _____ percent of their nameplate current.

15. Counter cooking tops and wall-mounted cooking units require special circuits and overcurrent protection in the form of fuses or breakers, usually _____ to _____ A

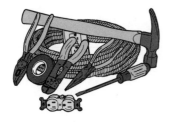

14
Light Commercial Wiring

_____   Score _____

_____ Period _____   Instructor _____

OBJECTIVE: You will be able to describe wiring for light commercial structures and multi-family or multiple occupancy structures.

DIRECTIONS: Carefully read Chapter 14 of the text. Then complete the following questions and problems.

Which of the following stores would most likely require only a simple 100 ampere service?

a. An appliance store.

b. A card shop.

c. An electronics store.

d. A heavy-duty tool shop.

A 30-unit apartment building may easily require a service entrance of:

a. 25 A.

b. 100 A.

c. 500 A or less.

d. 1000 A or more.

The lighting requirement for show windows in a small retail store is _____ VA per lineal foot.

The NEC sets the general lighting load at _____ VA per square foot for stores.

One and two-family dwellings are not considered multiple occupancy by the *Code.* (true or false)

In an apartment building, the electrical requirements of common areas, such as stairways, hallways, and outside lighting, are known as _____ loads.

Multiple occupancy dwellings of one or two stories must have main disconnects in each apartment. (true or false)

In three-story or greater multiple occupancies, _____ equipment and main disconnects for all apartments must be in a common, accessible area.

9. Describe a simple way to compute apartment loads.

   _____

   _____

   _____

   _____

10. Once individual apartment requirements are completed, you can then determine the main _____ _____ conductor sizes needed for the entire building.

15
Farm Wiring

e _____   Score _____

_____ Period _____   Instructor _____

ECTIVE: You will be able to describe the planning and layout considerations that take place when installing electrical equipment on a farm.

CTIONS: Carefully read Chapter 15 of the text. Then complete the following questions and problems.

A _____ should be thought of as a small industrial plant or business having several buildings, but involved in a single purpose.

At a farm, the service drop ends not at the dwelling, but rather at a centrally located _____ _____.

Almost all modern farms require a _____ to _____ A service entrance.

On a farm, the yard pole serves as a power center and overall disconnecting means for the farm. (true or false)

To insure proper grounding and safe wiring in farm buildings, which of the following should be used?
a. Nonmetallic cable.
b. Nonmetallic outlet boxes.
c. Metal armored cable.
d. Both a and b.
e. None of the above.

_____ - _____ light fixtures are often used on farms because of excessive damp and dusty conditions. The bulb is encased in a glass dome, around which is a strong, protective cage.

In farm buildings, receptacles should be mounted no lower than _____ inches (_____ m) above the floor. Higher is even better as it prevents them from being damaged by animals or tools.

In barns and other farm buildings, _____ lights direct the light downward where it is needed.

9. In a poultry house, _____ switches to increase egg production may be a part of the lighting circuits.

10. The farmhouse power requirements are figured the same as any residence. (true or false)

11. The farm building that will probably have the greatest electrical demand is the dairy barn. (true or false)

12. Power _____ of electric motors must be known to figure a motor's load demand on an electrical circuit.

13. The _____ _____ represents the power that would be needed if everything were operated at one time.

14. The _____ _____ represents the amount of power which would most probably be needed at any given time.

15. Generally, the minimum demand load is considered to be about _____ percent of the connected load.

16
Mobile Home Wiring

e _____

Score _____

_____ Period _____

Instructor _____

ECTIVE: You will be able to describe the special wiring requirements to meet the needs of mobile homes and mobile home parks.

CTIONS: Carefully read Chapter 16 of the text. Then complete the following questions and problems.

The _____ _____ is an assembly which includes bus bars, automatic over-current devices, and, sometimes, switches. This is enclosed in a cabinet or cutout box that is attached to the interior wall of the mobile home.

The conductors with their fittings and equipment, which carry electrical current from the mobile home service equipment to the distribution panelboard are called the _____ _____.

A single power cord with a molded plug meets *Code* requirements for a mobile home when the electrical load does not exceed _____ amperes.

An electrician will rarely be called upon to install a mobile home distribution panelboard since it is almost always supplied by the manufacturer of the mobile home. (true or false)

List three ways that an electrical feed can be supplied to a mobile home.

_____

_____

_____

A mobile home power supply cord is permitted if the load is _____ amperes or less.

The most common feeder assembly for a mobile home is:

a. A power supply cord.

b. An overhead installation.

c. An underground service lateral using cable or conduit.

d. None of the above.

8. Label the parts of the overhead feeder hookup for a mobile home.

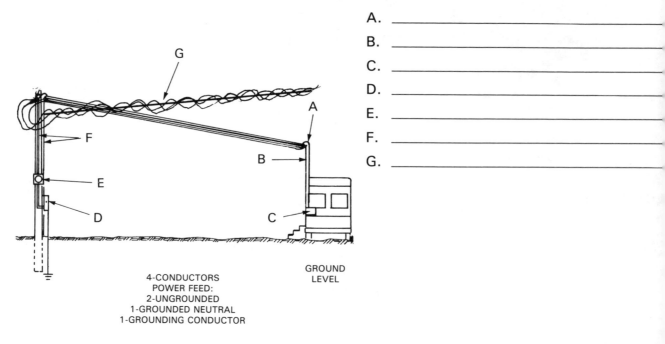

4-CONDUCTORS
POWER FEED:
2-UNGROUNDED
1-GROUNDED NEUTRAL
1-GROUNDING CONDUCTOR

GROUND
LEVEL

A. _____
B. _____
C. _____
D. _____
E. _____
F. _____
G. _____

9. Label the parts of the underground feeder hookup for a mobile home.

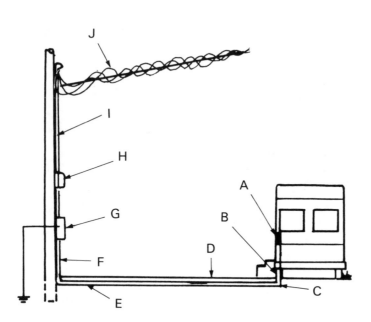

A. _____
B. _____
C. _____
D. _____
E. _____
F. _____
G. _____
H. _____
I. _____
J. _____

10. Service equipment must be mounted on the mobile home itself. (true or false)

Label the parts of the mobile home's distribution panelboard shown below.

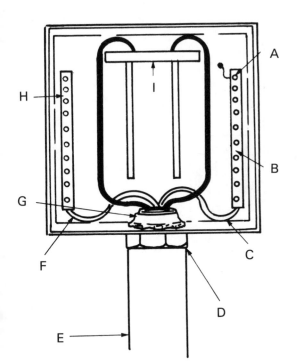

A. _____

B. _____

C. _____

D. _____

E. _____

F. _____

G. _____

H. _____

I. _____

In calculating panelboard load, what is the requirement for lighting and appliance circuits?

_____

_____

If a mobile home park is to accommodate 45 mobile homes, what would the amperage be for the park's main service entrance? (Show your work in the space provided.) _____ A

Common mobile home park facilities and equipment, such as security lighting, must be considered in the overall service demand. (true or false)

15. Label the parts indicated on the typical mobile home hookup shown below.

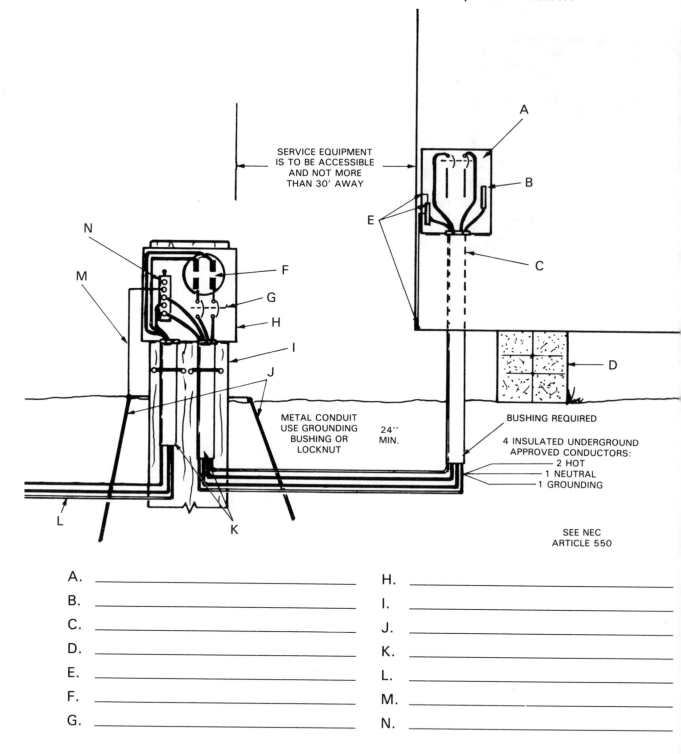

SERVICE EQUIPMENT
IS TO BE ACCESSIBLE
AND NOT MORE
THAN 30' AWAY

METAL CONDUIT
USE GROUNDING
BUSHING OR
LOCKNUT

24''
MIN.

BUSHING REQUIRED

4 INSULATED UNDERGROUND
APPROVED CONDUCTORS:
— 2 HOT
— 1 NEUTRAL
— 1 GROUNDING

SEE NEC
ARTICLE 550

A. _____   H. _____

B. _____   I. _____

C. _____   J. _____

D. _____   K. _____

E. _____   L. _____

F. _____   M. _____

G. _____   N. _____

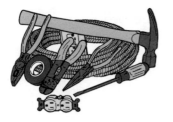

17
Low-Voltage Circuits

ne _____   Score _____

e _____ Period _____   Instructor _____

JECTIVE: You will be able to define remote control wiring and describe how the components operate. You will also be able to connect remote control switching systems.

ECTIONS: Carefully read Chapter 17 of the text. Then complete the following questions and problems.

Low-voltage circuits are those which operate at voltages much less than _____ V.

The job of a _____ _____, low-voltage circuit is simply to open or close a switch to start current flow in a higher voltage circuit.

Name the two basic parts of a relay.

_____

_____

Which of the following is a remote control (relay switching) and low-voltage circuit?

a. Intercom systems.

b. Chimes.

c. Lighting systems.

d. Fire alarms.

A _____ circuit is one which energizes devices specifically designed to give visual or audible signals.

List six advantages of remote control.

_____

_____

_____

_____

_____

_____

A low-voltage switching system operates at _____ volts of alternating current.

8. The heart of the remote control switching system is the step-down _____.

9. Which of the following is true about relays?

    a. The relay is usually installed in the fixture outlet box.

    b. The power operating the relay comes from the 24 V side of the transformer.

    c. Low-voltage leads are outside the box.

    d. The 120 V leads are inside the box.

    e. All of the above.

10. The low-voltage switches used in a remote control system are single-pole, double-throw types. (true or false)

11. A _____ _____ switch houses a number of remote switches, so any or all switches can be activited or deactivated from one location.

12. Low-voltage cable is very heavy and is very well insulated since the voltage is so low. (true or false)

13. Identify the devices represented by the low-voltage symbols shown below.

SYMBOL	DEVICE
— — – – —	
◇T◇	
◇R◇1 ◇R◇p	
◇S◇p ◇S◇t	
◇MS◇12	
◇MM B◇ ◇MM R◇	

14. Low-voltage wiring is not governed by the same rules as 120 V wiring. (true or false)

15. Before making low-voltage installations, check _____ _____ charts.

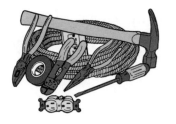

18
Electrical Remodeling

...e _____   Score _____

...e _____ Period _____   Instructor _____

...ECTIVE: You will be able to update and extend existing wiring systems.

...ECTIONS: Carefully read Chapter 18 of the text. Then complete the following questions and
problems.

Remodeling or modernizing electrical installation is perhaps the _____ (easiest, most
difficult) kind of wiring to do.

Special effort has to be taken to ensure the continuity and integrity of the _____ conductor when extending or adding to an existing system.

When remodeling wiring, why should you use fuses or circuit breakers rated for the amperage of
the circuit?

_____

_____

List four special tools that are needed when remodeling wiring.

_____

_____

_____

_____

Conduit is often used when remodeling wiring, especially in finished walls. (true or false)

When making wall openings for switch boxes and outlet boxes and holes for wire-pulling, the
openings should be made as _____ (large, small) and as neat as possible.

_____ _____ are a practical necessity for running cable in old work.

When fastening cable inside a nonmetallic box, at least 1/4 inch of cable jacket should be pulled
through the clamp. (true or false)

When extending an outlet, what two factors should be considered?

_____

_____

10. If you are installing a ceiling fixture, why would going through the attic space be the most obvious route?

_____

_____

11. Baseboard installations can be done without the use of a fish tape. (true or false)

12. Before working on an outlet, a _____ tester should be used to make sure that the power is off.

13. Look at the illustration below. Identify four methods of routing wiring during remodeling.

 A. _____

 B. _____

 C. _____

 D. _____

ADDING ON

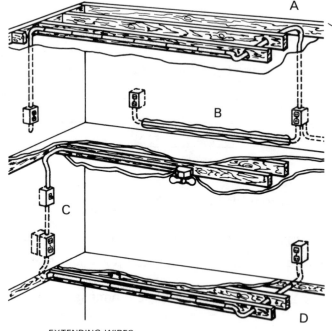

EXTENDING WIRES

4. A box must be prepared and the cable attached to it _____ (before, after) it is installed.

5. It is best to replace old, outdated electrical equipment because overloaded circuits and extension cords are _____ hazards.

6. When installing a sub-panel, it is usually mounted as near to the _____ _____ as possible.

7. Identify the items indicated below.

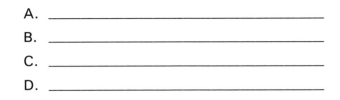

A. _____

B. _____

C. _____

D. _____

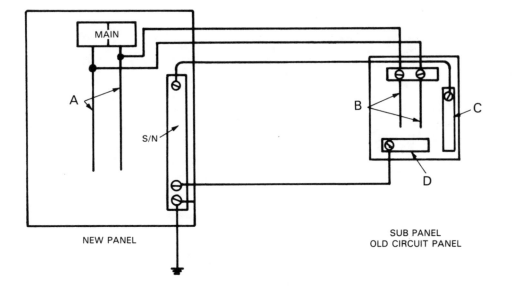

8. A surface _____ is a simple, inexpensive way to extend a circuit to get more receptacles.

19. What is the main benefit of a surface wiring system.

_____

_____

20. When installing surface wiring, a typical installation begins with bringing the line side conduc tors into the base of the surface assembly. (true or false)

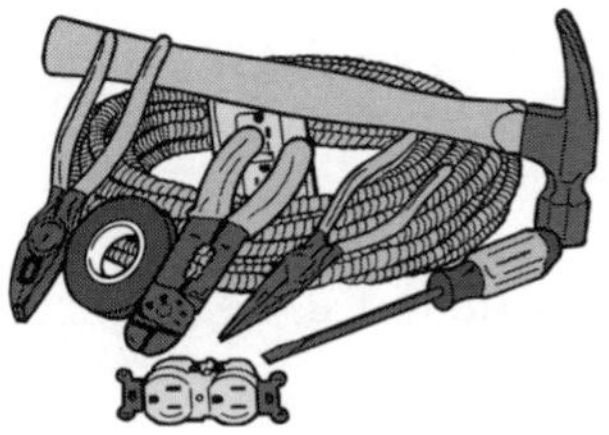

19
Electrical Meters

ne _____

e _____ Period _____

Score _____

Instructor_____

JECTIVE: You will be able to identify and use electrical meters.

ECTIONS: Carefully read Chapter 19 of the text. Then complete the following questions and problems.

. The basic mechanism of electrical meters like the ammeter and voltmeter is the _____ movement.

. List six types of meters that an electrician would use.

_____

_____

_____

_____

_____

_____

. An _____ is used to measure the current flow in an electric circuit.

. The _____ is an electrical instrument used to measure the difference in potential or the voltage across two points along a circuit.

. The _____ is used to measure circuit resistance.

. Identify and describe the instrument shown below.

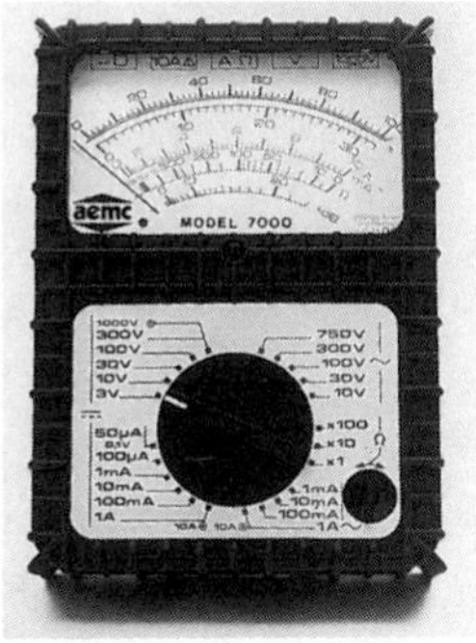

(AEMC Instruments)

_____

_____

7. By adding one or several _____ into the meter circuit, a dc meter can be converted into one which will measure ac electricity.

8. A meter designed for measuring power in an electric circuit is called a _____.

9. The kilowatt hour meter below would be read as _____ kW/hr.

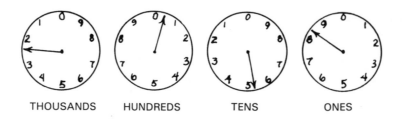

THOUSANDS   HUNDREDS   TENS   ONES

10. List six guidelines to follow when using and caring for meters.

_____

_____

_____

_____

_____

_____

20
Electrical Troubleshooting

Score _____

Period _____ Instructor _____

ECTIVE: You will be able to trace common troubles in an electrical system through troubleshooting.

CTIONS: Carefully read Chapter 20 of the text. Then complete the following questions and problems.

_____ must be uppermost in the troubleshooter's mind.

Name and describe four instruments that can be used in troubleshooting.

_____
_____
_____
_____
_____
_____
_____
_____
_____
_____

If no voltage is present at the receptacle slots, but a positive test lead reading is obtained, the chances are very good that the _____ is open.

Explain how to check the ground continuity on an older type receptacle that does not have a ground slot.

_____
_____
_____

To test a switch, remove the cover plate and determine if there is power to the switch by touching one lead of the neon tester to the metal box and the other lead to the line side terminal. If the tester lights, the circuit _____ (is, is not) working.

The substitution of a single three-wire cable for two two-wire cables constitutes a multiwire branch circuit. (true or false)

7. The most common problem occurring with split or multiwired circuits is the improper connection of the circuit conductors at the _____.

8. If a fuse "blows," it should be replaced with one having a higher rating. (true or false)

9. List two ways in which you can identify a blown cartridge fuse.

   _____

   _____

   _____

   _____

10. If a breaker "trips":
    a. Find the cause before resetting the breaker.
    b. Reset the breaker and ignore the cause of the "trip."
    c. Never reset the breaker again.
    d. None of the above.

21
Specialized Wiring

Score _____

Instructor _____

:CTIVE: You will be able to describe and perform specialized electrical wiring installations.

CTIONS: Carefully read Chapter 21 of the text. Then complete the following questions and problems.

_____ circuits supply electrical energy to doorbells, buzzers, signal lights, and other warning devices.

Conductors serving buzzers, bells, or chimes must never be run with regular or full-power circuits that operate on 120 or 240 V. (true or false)

Doorbell, chime, or buzzer transformers operate at:

a. 6 to 20 volts.

b. 40 to 80 volts.

c. 120 volts.

d. 240 volts.

List and describe the three classes of hazardous locations as determined by the NEC.

_____
_____
_____
_____
_____
_____

List and describe the two divisions of hazardous locations.

_____
_____
_____
_____

6. A factory that handles or transfers flammable or explosive gases would be a:
   a. Class I, Division 1 hazardous location.
   b. Class II, Division 1 hazardous location.
   c. Class II, Division 2 hazardous location.
   d. Class III, Division 2 hazardous location.

7. Which of the following hazardous locations should be wired by methods which are sealed to keep out dust, lint, and fibers?
   a. Class I, Division 2 hazardous locations.
   b. Class II, Division 1 hazardous locations.
   c. Class II, Division 2 hazardous locations.
   d. Class III, Division 2 hazardous locations.

8. Facilities such as service stations and aircraft hangers often fall under more than one class and division of the hazardous locations classifying scheme. (true or false)

9. In most cases, a garage or outbuilding should have circuits which are separate from the main structure. (true or false)

10. Name two alternate types of wiring systems that take over when the regular electrical power source is not functioning as identified by the NEC.

   _____

   _____

11. Emergency systems are intended to automatically supply enough power to assure the safety of a building's occupants during a _____ _____.

12. Emergency systems are required mostly in places such as:
   a. Single-family residences.
   b. Buildings or building services whose failure to operate could produce a serious threat to life.
   c. Barns.
   d. Garages or other outbuildings.

13. Emergency systems must be _____ and _____.

14. The transfer switch for emergency systems should be located _____ (ahead of, behind) the main service disconnect.

15. The NEC requires that wiring from an emergency source and overcurrent protection to emergency loads be kept:

   a. Completely separate from all other wiring and equipment.

   b. Connected to all other wiring and equipment.

   c. Connected to some other wiring and equipment.

   d. None of the above.

16. Standby systems serve loads necessary to the normal operation of vital building systems. (true or false)

17. Standby power is legally required in residences and private businesses. (true or false)

18. Emergency power equipment is almost always tied to a gasoline or diesel engine powered _____.

19. Telephone equipment is manufactured and sold by many different commercial firms. (true and false)

20. Only telephone company employees can install telephone equipment. (true or false)

21. List two types of telephone systems.

   _____

   _____

22. There is little danger of electrical shock from the telephone voltage. (true or false)

23. List the three kinds of indoor telephone wiring.

   _____

   _____

   _____

24. When attaching cable pairs, leave some _____ in the wires so the wires will not pull loose from the block.

25. Most terminal blocks allow you to run cable pairs for as many as _____ phones.

26. Because of the lower voltages found in telephone systems, _____ are not required at junctions.

27. Why is the fish tape useful in telephone installation?

_____

_____

28. The _____ _____ is the jack or terminal where the telephone company installation terminates.

29. No more than _____ phones can be installed without overloading the incoming phone circuit.

30. In new construction, the telephone wiring should be installed _____ (before, after) the application of drywall or paneling.

31. Care should be taken to keep telephone wiring _____ (close to, away from) electrical wires and electrical devices.

32. Installing telephone systems in old work is the easiest type of telephone installation. (true or false)

33. Telephone wires can be made less noticeable by placing them:
    a. In areas where moldings or grooves in paneling partially or wholly conceal them.
    b. Inside cabinets.
    c. Under carpet.
    d. Both a and b.

34. List the colors of the four wires of telephone cable and describe their common purposes.

_____

_____

_____

_____

_____

_____

_____

_____

35. If you suspect an electrical short in a phone system, use an _____ to find the problem.

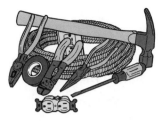

22
Motors and Motor Circuits

ne _____   Score _____

e _____ Period _____   Instructor _____

JECTIVE: You will be able to describe the wiring requirements for electric motors. You will also be able to order the right replacement when an old motor is no longer available.

ECTIONS: Carefully read Chapter 22 of the text. Then complete the following questions and problems.

. Motors with a higher voltage than 600 V, such as those found in larger commercial and industrial operations, should be handled only by maintenance electricians having considerable experience with these motors. (true or false)

. The motor _____ gives motor characteristics.

. List the information that is required to appear on the nameplates of all motors.

_____

_____

_____

_____

_____

_____

_____

_____

_____

_____

_____

_____

_____

_____

_____

_____

_____

4. The higher the locked-rotor kilovolt-ampere (kVa) is, the _____ (higher, lower) the starting current surge will be.

5. For standardization, NEMA has assigned the _____ size to be used for each integral horsepower motor so that shaft heights and dimensions will be the same to allow motors to be interchanged.

6. If applied voltage varies too much from the nameplate specifications, it will produce noticeable changes in the motor _____.

7. The greatest allowable voltage drop on a motor circuit is basically related to the _____ and _____ of a motor.

8. List four common motor circuits.

   _____

   _____

   _____

   _____

9. For a single motor circuit, the NEC requires that the feeder capacity be _____ percent of the full-load current rating of the motor.

10. The maximum fuse or breaker size permitted for short-circuit or ground fault protection is _____ percent of the full-load current rating.

11. List four requirements that a single motor branch circuit must provide.

    _____

    _____

    _____

    _____

12. When each motor is 1 hp or 6 A or less, the circuit will require a _____ A or smaller fuse or breaker.

13. In a motor branch circuit, each motor needs motor running overcurrent protection, and this protection must not exceed the amperage stamped on the ''approved for group installation'' overcurrent device of the _____ (smallest, largest) motor on the branch circuit.

14. The NEC defines a motor _____ as a switch or other device normally used for starting and stopping a motor.

15. On small, single-phase motors, the overcurrent device may serve as the controller for motors of _____ hp or less.

16. A general-use type switch may serve as the controller for motors up to _____ hp.

17. A cord and plug may serve as the controller for motors having less than _____ hp.

18. All electric motors are designed for:

    a. Continuous duty only.

    b. Limited duty only.

    c. Either continuous duty or limited duty.

    d. None of the above.

19. Define rated-load current.

    _____

    _____

    _____

20. The disconnecting means rating for a hermetic motor shall be 115 percent of the nameplate rated-load circuit or the branch-circuit selection current, whichever is _____ (less, greater).

21. List five sizing considerations that must be determined in order to properly install a circuit for combination loads.

    _____

    _____

    _____

    _____

    _____

22. The maximum (hot spot) continuous temperature of 130 °C (266 °F) for small induction motors would be a:

    a. Class A system.

    b. Class B system.

    c. Class F system.

    d. Class H system.

23. Normal maximum ambient temperature is _____ °C (_____ °F) for most motor ratings.

24. Preventive maintenance and proper motor loading are the best insurance against motor _____.

25. List four causes of motor failure.

    _____

    _____

    _____

    _____

26. List four factors that can cause a motor to overheat.

_____

_____

_____

_____

27. _____ may fail (seize) in unused motors that are not rotated for extended periods.

28. Secure mounting and correct _____ with the load are essential for proper motor performance.

29. List three ways motors may be connected to a load.

_____

_____

_____

30. If a loss of power occurs, a time-delay relay in the magnetic starter control will provide for random restarting to prevent the excessive voltage drop in the wiring that would occur if all the motors came on at one time. (true or false)

23
Swimming Pool Wiring

Score _____

Instructor_____

OBJECTIVE: You will be able to describe wiring considerations related to swimming pools.

DIRECTIONS: Carefully read Chapter 23 of the text. Then complete the following questions and problems.

1. Name four swimming pool components that must be bonded using a No. 4 AWG solid copper conductor.

_____

_____

_____

_____

2. On the illustration of the typical swimming pool below, identify those components that need to be bonded.

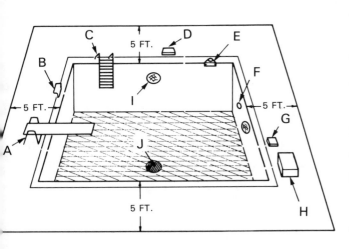

A. _____

B. _____

C. _____

D. _____

E. _____

F. _____

G. _____

H. _____

I. _____

J. _____

3. In addition to bonding, all electrical equipment related to the pool and within 5 ft. of the pool walls must be properly _____.

4. With the exception of a locking type receptacle for a cord connected swimming pool pump, *Code* does not permit any receptacles to be located within _____ ft. of the pool walls

5. Any and all switches or switching devices of any type must be set back at least _____ ft. from the pool edge.

6. List the three categories of pool lighting.

   _____

   _____

   _____

7. List six requirements underwater lighting fixtures must meet in addition to grounding, bonding and sealing.

   _____

   _____

   _____

   _____

   _____

   _____

8. Name the two types of fixtures installed below the waterline in pools.

   _____

   _____

9. Overhead utility conductors must not be within _____ ft. of the horizontal perimeter of the pool.

10. True or false? The number one priority concerning all electrical items to be wired within or around a swimming pool is to assure proper grounding and bonding.

24
Electrical Careers

Name _____

Date _____ Period _____

Score _____

Instructor _____

OBJECTIVE: You will be able to identify and describe various careers in the field of electrical wiring.

DIRECTIONS: Carefully read Chapter 24 of the text. Then complete the following questions and problems.

1. List five types of electrical careers.

_____

_____

_____

_____

_____

2. Line _____ and _____ construct and maintain the network of power lines that carries electricity from generating plants to consumers.

3. _____ are experienced line installers and repairers who are assigned to special crews that handle emergency calls.

4. When cables are installed, the _____ _____ pull the cable through the conduit and then join the cables at connecting points in the transmission and distribution systems.

5. Which of the following do maintenance electricians keep in good working order?

   a. Lighting systems.
   b. Transformers.
   c. Generators.
   d. All of the above.

6. An _____ usually lasts four years and consists of on-the-job training and related classroom instruction in subjects such as mathematics, electrical and electronic theory, and blueprint reading.

7. Many people learn the electrical trade informally on-the-job by serving as _____ to skilled maintenance electricians.

8. All electricians must be familiar with the _____ _____ _____ and local building codes.

9. Heat, light, power, air conditioning, and refrigeration components all operate through electrical systems that are assembled, installed, and wired by _____ electricians.

10. List seven high school or vocational school courses that someone considering a career in electronics should take.

_____

_____

_____

_____

_____

_____

_____

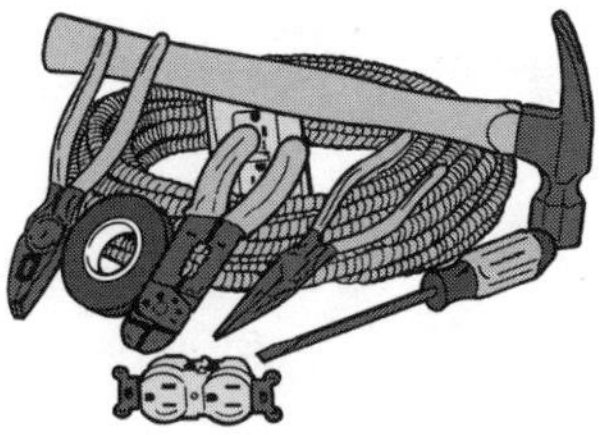

Math Review

ne _____   Score _____

e _____ Period _____   Instructor _____

OBJECTIVE: You will be able to perform math operations used by the electrician.

DIRECTIONS: Carefully read Chapter 25 of the text. Then complete the following questions and problems.

Explain why math must be considered as an important electrical tool.

_____

_____

_____

Change 5/16 to a decimal. (Show your work in the space provided.)

_____

Change 0.1875 to a fraction. (Show your work in the space provided).

_____

What percentage of 89 is 12? (Show your work in the space provided.)

_____%

5. What percentage is the fraction 3/4? (Show your work in the space provided.)
   _____%

6. What is the Ohm's law formula?
   _____

7. When using Ohm's law, if you wish to find I when E and R are known, how is the formula rewritten?
   _____

8. How is trigonometry used in electrical work?
   _____
   _____
   _____

9. What is the formula used to obtain the area of a circle?
   _____

10. The area of a rectangle is:
    a. Any side times itself.
    b. Equal to its base times its height.
    c. Found by adding its four sides.
    d. Four times any side.